Environmental Security and India

This book examines environmental issues through the lens of security studies and presents a comprehensive analysis of Indian policy in dealing with threats posed by climate change.

This book

- Puts forward theoretical base for securitization of environmental issues, incorporating different schools of thought;
- Presents a survey of global environmental politics in general and the effects of climate change and its consequences for India's national security in particular;
- Examines the politics involved in India's environmental policy at both the domestic and international levels;
- Outlines key policy takeaways and possibilities for action that can help contain the threat of environmental change.

A comprehensive guide to a new and emerging dimension in Indian security policy, this book will be essential reading for students and researchers of international relations; security studies, especially non-traditional security; public policy, especially environmental policy; and area studies.

Satabdi Das is Assistant Professor in the Department of Political Science, South Calcutta Girls' College, India. She has completed her Ph.D. in *Environmental Security and India: Global Concerns and National Interest* at the Department of International Relations, Jadavpur University, India. She has publications with several major publishers and in UGC-enlisted peer-reviewed journals. She has presented papers at many national and international conferences organized within the country and abroad. Her paper on *Sustainability Education in Agriculture: The Prospect of Digital Green in India* was considered among the first five papers in the International Conference on Sustainability Education organized by the Mobius Foundation, the Climate Reality Project and UNESCO in New Delhi, during 9–10 September 2019. Presently, she is working on sustainable development, environmental movements and climate politics.

Environmental Security and India

Global Concerns and National Interest

Satabdi Das

Routledge
Taylor & Francis Group

LONDON AND NEW YORK

First published 2023
by Routledge
4 Park Square, Milton Park, Abingdon, Oxon OX14 4RN

and by Routledge
605 Third Avenue, New York, NY 10158

Routledge is an imprint of the Taylor & Francis Group, an informa business

© 2023 Satabdi Das

British Library Cataloguing-in-Publication Data
A catalogue record for this book is available from the British Library

Library of Congress Cataloging-in-Publication Data
A catalog record has been requested for this book

ISBN: 978-1-032-20586-1 (hbk)
ISBN: 978-1-032-22111-3 (pbk)
ISBN: 978-1-003-27119-2 (ebk)

DOI: 10.4324/9781003271192

Typeset in Sabon
by Apex CoVantage, LLC

This book is dedicated to my parents, my husband and my daughter for their endless love, support and encouragement

Contents

Figures

Tables

Acknowledgements

First and foremost, I would like to sincerely thank my supervisor, Professor Dr. Anindya Jyoti Majumdar, Department of International Relations for his guidance and support throughout this study. I would like to acknowledge the Department of International Relations, Jadavpur University, the departmental library, the editorial board of JJIR, the Department of Geography, Mizoram University, for providing me with support, guidance and other academic inputs.

I have to thank my parents late Mr. Dilip Kumar Das and Mrs. Siuli Das and my sister Piyali Basu for their love and support throughout my life. I am also grateful to my husband Dr. Jyoti Prasad Saha and my daughter Gunja for their immense support and sacrifice during the entire period of my work. My other family members especially my mother-in-law Kanan Saha and father-in-law Nandadulal Saha deserve my whole-hearted thanks.

Introduction

Issues of 'Environmental Security' affect not only the geographical landscape of a country but its economy and socio-political structure too. Further, as environmental threats transcend national borders, unilateral state action is insufficient to deal with them. Various efforts are therefore made at all levels – national, regional and international, to conserve our surroundings and to make the path of development more sustainable to promote peace, stability and human security. The problem is now a matter of global concern too. Both the academia and the policymakers are therefore concentrating on the necessity of extending our conventional understanding of the notion of security to incorporate environmental issues like scarcity of renewable and non-renewable resources, environmental degradation as well as climate disasters and their effects on life and livelihood of people across the world. Today, these issues have introduced a new agenda in the discourse of security studies.

Security is such a term which underscores the absence of threats to one's well-being. Whenever this term is used, a set of questions arise like whose security, what are the possible means to achieve it and how long we will be able to preserve the conditions required to sustain security. Barry Buzan's work is a standard source for many recent discussions on the theme of security in international relations. He candidly put this in the following way:

> the foreign, economic and military policies of states, the intersection of these policies in areas of change or dispute and the general structure of relations which they create are all analyzed in terms of national and international security.[1]

However, the essence and connotations of the term change with time. During the cold war days as the world lived on the 'brink of annihilation' facing the threat of a clash between two nuclear superpowers, the notion of security was then viewed through the prism of interstate struggle for power. So traditionally security is analysed in terms of military dimension. In other words, the idea of security was then limited to parochial and conceptual frameworks of national security, particularly determined by the parameters

DOI: 10.4324/9781003271192-1

of 'dominance' and 'stability', impinging upon the traditional concept of the maintenance of balance of power.

However, with the end of the cold war, the threat agenda we face requires a more holistic approach to non-traditional and non-state-centric security threats. In a bipolar world structure, the capacities and intentions of the enemies were, if not confirmable, at least comprehensible.[2] But the perturbations and turmoil across the world created by such new threats like poverty, human trafficking, refugee influx, the spread of organized crime and many more cannot be now checked through military means only. Securitization of these issues, which traverse national borders and are posited beyond the realm of interstate military conflict, has given birth to many scholarly debates and theory building since the 1980s. Early debates on this topic, basically pioneered by the traditional realists, are largely determined by the belief that widening the security agenda would risk making both the scholarship and state policy incoherent. But there are other studies which consider such widening is necessary and emphasize the inclusion of human security issues.[3] The Copenhagen School of thought on security, which was articulated by Barry Buzan et al., talked about the various processes by which any issue may pose an existential threat if it is looked through the prism of securitization process. They also made efforts to link up security to five obvious sectors. These sectors are military sector, political sector, economic sector, societal sector and the environmental sector.[4] So security cannot be defined in terms of military preparation for protecting the territorial integrity of nation states alone. Non-traditional security issues have thus come into prominence in recent years to rebuild the contours of security discourse. They have altered the tendency to consider some issues from the point of view of national concerns only.[5] So in the present epoch the scope and parameter of threats have expanded which blur national borders. With this change in security dynamics, the security lexicon is also broadened in order to incorporate various non-military considerations.

Environmental degradation and climate change is one of those issues that have entered the realm of security discourse today. It is quite evident that human intervention in the planetary atmosphere has caused severe detriments to human lives. Buzan et al., while talking about several sectors with which security is connected quite aptly, suggested that the sectors can be understood with reference to specific types of interaction. To them, environmental sector is about relationships between human activity and the planetary biosphere. In this sense, environmental security 'concerns the maintenance of the local and the planetary biosphere as the essential support system on which all other human enterprises depend'.[6] Thus, a shift of focus is visible in the concept of security from the cold war and realist preoccupations with statist notion of security to people-oriented approach to security where the referent object is not the state but the human beings.

The term environmental security has many connotations. There is a common trend to use the terms – environmental security and environmental protection – interchangeably. Therefore, any effort to extend the notion of security to the domain of environment has met with debates and differences in opinions in academic and policy arena. It may denote ensuring security against natural- or human-induced environmental dangers, protecting the environment from military actions or restoring peace in the face of a conflict that took place due to competing claims of finite environmental resources. It is at that juncture, very difficult to reach an all-encompassing definition for environmental security. The causes, manifestations, intensity and consequences of environmental degradation are diverse. Countries' vulnerability to it and their coping strategies are also different. Additionally, environmental security is inextricably linked with economic and social environment of a particular country which constitutes the political system. It is the socio-political setting of the country that actually determines the fair accessibility of individuals to environmental goods and services. Nonaccessibility to these goods ultimately leads to conflicting situations and civil and international strife in some regions. All such conflicts are not directly emanating from environmental scarcity only but instigated by unjust distribution policies too. The rising gap between the rich and the poor countries in terms of resource consumptions has intensified the problem further. Developing countries are more environmentally insecure as most of their populations are dependent on climate-sensitive sectors for their lives and livelihoods and their economic, social and political institutions are already fragile and plagued with hunger, disease, abject poverty, unrestrained population growth and so on. Such grim situations are the results of imbalance in economic and social structures. However, the dichotomy is not limited to rich and poor countries of the world but between rich and poor masses of the poor impoverished southern countries as well.

Against this backdrop, in this study, environmental security is interpreted in terms of a situation where natural resources like crops, lands, water, forest and energy resources are protected from natural- and human-induced environmental dangers, and people have the required measures at their disposal to combat these threats. Environmental degradation, global warming and climate change are not security threats in any conventional sense. If the term security is defined in a broader horizon to incorporate human, social, physical and economic well-being of a nation, then environmental distress acts as a risk amplifier that exacerbates existing animosities between communities in some regions and worsens the already strained socio-economic and political situations of people living at the edge. In this sense, environmental security is related to human security too. Actually, environmental degradation and climate change have affected various dimensions of human security. It is a multifaceted concept. The Human Development Reports (HDR) of the UN in its 1994 issue, while enumerating seven elements of human security, included environmental security as one of the dimensions that is inextricably linked with human security. Other elements of it like economic, food and health security are also hampered due to environmental change induced 'loud and silent emergencies'.[7] Here, human security implies the assurance of overall human well-being that is at stake due to environmental assaults. Security in the traditional territorial sense cannot solve these problems and environmental insecurity as a non-conventional threat requires actors and institutions beyond the ambit of state system as ecological strains spill beyond national frontiers.

However, environmental problems can be both natural and human-induced. In the anthropocentric era, we are interfering in the realm of natural environment like diverting the natural riverscape, denuding forest cover, unscrupulously extracting the groundwater, changing land use pattern for business use and many more. All of these have led to the destruction of our natural environment. Such re-engineering in the environmental landscape in one country may cause havoc in other countries.

Given such transnational impacts of natural and man-made environmental change, it is in the best interest of all nations to discern ways of cooperation to combat climate threats and its various manifestations. Along with concern for climate-driven resource scarcity, extreme weather events and population displacements, countries have tried to deal with the direct cause of environmental change that is emissions of Greenhouse Gases (GHGs) and the resultant global warming. Therefore, policies are devised time and again to reduce GHGs and move towards a low-carbon trajectory through the imposition of various regulations and incentives. But as the threats to climate security are generally transnational in nature, their resolution requires commitments and coordination from a wide range of actors – at national, regional and global levels. It is not very easy to agree on cross-border and global environmental issues when the ecological concerns and interests of the countries differ. Therefore, a different kind of politics has evolved where multiple stakeholders are negotiating and bargaining an environmental regime. In this whole process, they are trying to safeguard their respective national interests which have made an all-encompassing environmental deal intractable.

Against this backdrop, in India, environmental change has been turned into a major non-traditional security concern instead of remaining only an environmental issue. Many environmental issues like resource scarcity, loss of biodiversity, extreme weather events, sea level rise and the resultant behavioural strategy choice of migration – all are happening here requiring political attention. These are threats to the entire economic political and social well-being of the masses. To strengthen India's national capacities for environmental management, monitoring systems, planning capacities, personnel and resources for implementing plans should be improved. But to minimize environmental threats, the risks that are the results of environmental change have to be looked at in the context of interdependence. Our systems for environmental management are basically national. But border controls are not possible options for managing transboundary pollution or impact of carbon emissions. In various Conference of Parties, cooperation among nations has been proved vitally important. India in these conferences tried to raise her voice to protect her national interest vis-à-vis the global concerns for the threat. Here, national interest does not imply the goal of enhancing military security. Rather states are here considered as the prime actors in fostering all sorts of human security including the environmental one. India's negotiating position thus at different global environmental parleys is determined by her national interest to provide all its citizens' environmental security which is intertwined with her developmental aspirations and politico-economic and social security too.

As part of wider political process and debate, environmental issues have become an integral part of our existence. Therefore, the objective of the present study is to highlight the environmental risks that the global community has faced today and how particularly India is affected by them. The dangers of these risks are real and severe. While they have amplified the existing socio-political tensions of any given society, their intensity has become more acute due to the urge for unlimited economic growth and developmental aspirations of some countries as well. So the study concentrates on the environmental dimension of geopolitics in general and the effects of climate change and its consequences for India's national security in particular. It also aims to evaluate the stated positions of major actors in the environmental politics of international relations and to discern the politics involved in India's environmental policy both domestic and international – more specifically, her perceptions of the problem of environmental change and the kind of negotiating position that follows from these perceptions; the policies India has undertaken so far and finally India's possibilities for action that can help contain the threat of environmental change.

Notes

1 Barry Buzan, *People, State and Fear: The National Security Problem in International Relations*, Harvester Press: Great Britain, 1987, p.3.
2 Ronnie D. Lipschutz (ed.), *On Security*, Columbia University Press: New York, 1998, p.1.
3 Shahar Hameiri and Lee Jones, "The Politics and Governance of Non-Traditional Security", *International Studies Quarterly*, Vol.57, No.3, September 2013, p.463.
4 Simon Dalby, *Geopolitical Change and Contemporary Security Studies: Contextualizing the Human Security Agenda*, Working Paper No. 30, Institute of International Relations, The University of British Columbia, April 2000, p.4. Available at www.cir.ubc.ca/download/i/mark_dl/u/4006903219/.../webwp30.pdf Accessed on December 28, 2014.
5 Shahar Hameiri and Lee Jones, No.3, p.465.
6 Bary Buzan, Ole Waever and Jaap de Wilde, *Security: A New Framework for Analysis*, Lynne Rienner: Boulder, CO, 1998, p.8.
7 UNDP, *Human Development Report*, Oxford University Press: New York, 1994, pp.24–25. Available at http://hdrundp.org/sites/default/files/reports/255/hdr_1994_en_complete_nostats.pdf Accessed on August 23, 2014.

1 Global Environmental Risks and Their Various Manifestations

Global environmental change presents a new and very different pattern of national security challenge today which requires non-state and non-military policies and strategies. The increased use of natural resources and their degradation, the rising use of fossil fuels and industrial pollution leads to the accumulation of excessive Greenhouse Gases (GHGs) in our surroundings. In coming decades, this phenomenon may overstretch many nations' adaptive capabilities that could in turn result in new political, economic and geophysical challenges ultimately jeopardizing national and international security interests of countries. The transnational nature and all-pervasive, intergenerational implications of environmental change have influenced nation states to consider it not only as a sole scientific phenomenon but also as an issue of both global agenda for peace and security and national policy concern.

Rising temperature, glacial melting and natural disasters result in the fact that some nations may have impaired access to water and food crops, dirty energy resources may increase the pollution level, forest covers are denuded affecting both the biodiversity and livelihoods of forest-based communities, rising sea levels may threaten the existence of coastal cities leading to the uproot of large number of people and so on. Environmental change has directly thus affected various regions. Competing claims over diminishing resources that have interacted with severe discrimination in distribution of these resources and with lopsided developmental policies in some societies have worsened the situation. They have also exacerbated the existing ethnic and religious conflicts among various communities in vulnerable regions of the world. Often, environmental migrants cross borders in search of resources which may create conflictual relations between the original residents of the host country and migrants too. Against such a backdrop, environmental change has been regarded as a threat multiplier affecting the living conditions in many Asian, African and Middle Eastern nations. Their national interests which are aiming at ensuring the human, social, physical and economic well-being of the population are thus also affected.

DOI: 10.4324/9781003271192-2

The concern for environmental security has grown rapidly since long, and its growing significance was evident in many major environmental conferences and symposia. As a result, various conventions and landmark protocols related to environmental protection came into prominence. The Brundtland Report (1987) which is named 'Our Common Future' also asserts the responsibility of environmental change behind the conflictual relationship between and among nation states. It states that political tension and military conflicts may emerge due to environmental stress. There are instances where nations in order to assist or resist control over several environmental resources have come to battle. As these resources became scarcer, such conflicts are likely to increase.[1] Same is echoed in Van Creveld's writing where he underscored that environmental scarcity might give birth to many future wars that would be of communal survival, aggravated by it. He predicted that such wars would be sub-national, implying that states and local governments would find it difficult to protect their own citizens physically which might lead to their end in many cases.[2] Basically, climate change triggered some of the impacts like scarcity of natural resources, which require domestic actions first to manage, whereas the effects of extreme weather events, paucity of water in the shared rivers, sea level rise go far beyond territorial borders. The effects of latter events also exacerbate the unavailability of natural resources. These situations are worsened due to population pressure and its resultant rising per capita consumption of resources and because of harmful process of extracting the resources. All such threats have necessitated international frameworks and engagements for actions. But at both global and individual country levels the necessary political will to act in unison is absent which has contributed to the rising intensity of this problem. The following sections reflect how environmental change poses risk to human existence in various parts of the world.

1. Resource Scarcity

When climates change significantly for which necessary resources like croplands, forests and water supply are not properly available to many, the situation leads to existential threats to many societies. Thomas Homer Dixon said that in regions where most of the population is more dependent on renewable, they are depleted faster than their replenishment. Evidence are stark in places ranging from Gaza to the Philippines to Honduras. As a result, there are instances of overdrawing and salinization of aquifers, disappearance of coastal fisheries, forest denudation and many more.[3] Thus, some major areas can be identified where environmental degradation and climate change have the potential to cause critical negative developments like the reduction of arable lands, impairment of food production, water scarcity and increasing natural calamities. As an indirect result of these developments, environmental change induced migration may increase in many fragile regions of the world.

1.1. *Food Scarcity and Land Degradation*

Food security exists if all of the population can have access to enough quantity of food which is safe and nutritious also, which can be procured by all irrespective of rich and poor as well as which are required for maintaining a healthy life.[4] But often the access, availability, utilization and stability of food are severely affected by climate variability. Temperature increase and changing patterns of seasons due to global warming may lead to heat stress for plants ultimately resulting in lowering production. The rising sea levels and unavailability of water resources may also affect food productivity. Additionally, environmental change affects food utilization through its impact on vectors of pests and diseases, and it also influences the availability of food through the escalation of recurrent natural disasters.[5]

All of these have brought new challenges for every nation which is concerned with national food security. The growing demand for food in such a situation has made the problem more acute as altered pattern of food production is related with the socio-economic condition and political setting of a given society which is stratified according to income level and accumulated wealth of the inhabitants. The pattern of demand for food is determined by this level of income and the process of distribution which is unjust and unequal mostly in developing countries. As food insecurity is closely linked with the insecure supply and the low accessibility of food, the challenge of unavailability of this resource due to famine and climate-induced low productivity is exacerbated by the supply-side constraints. This is the reason behind the fact that despite the rise in food production and reserves in many countries due to the Green Revolution, the number of hungry people did not decline. The declining yield may lead to food price hike that also has jeopardized the fundamental need of any society to feed her population. Besides, when the people of a country are spending much of their income on food, they become more vulnerable to other stressors.

Gradually, the severity of threats to natural resources is becoming a matter of concern for global community as they have the potential to fuel existing conflict over depleting resources including croplands, forests, food and water supply, specifically where their availability is highly politicized. At this juncture, environmental change may act as a threat multiplier. Climate extremes like abnormal temperature rise and changing pattern of precipitation and the consequent effects like droughts, floods, heatwaves and extreme cold have affected the availability and stability of food to a large extent. However, in some higher latitude regions, like in Russia and Northern America, global warming is conducive to grain production while it may change environmental condition to such an extent for which significant reduction in agricultural production is faced by some tropical countries. The agricultural impacts of climate change thus differ both within and across regions because of varied soil characteristics, regional climates, agro-economic conditions and so on.[6] During the period 2011–2016, severe droughts hit the world, leading to crisis-level food insecurity for 124 million people in 51 countries. The El Niño event of 2015–2016, which was exacerbated by climate variability, affected the dry corridor of El Salvador, Guatemala and Honduras which experienced one of the worst droughts in the past 10 years, affecting 50–90 per cent of the crop harvest. In the period of 2007–2010, Pakistan witnessed severe climate disasters having adverse effects on the agricultural sector.[7]

The climate change driven food crisis affects the poor masses of the tropics more because of their high exposure to natural disasters and direct dependence on climate-sensitive resources like plants, water, land and above all because of their least adaptive capacity to climate change. Since 1970s, drought and desertification affected the North African region immensely. According to an estimation by UNEP, the boundary between semi-desert and desert in Sudan has moved southward hugely. People were forced to move southward leading to struggle between farmers and herders.[8] Because of the strained food supply due to such extreme weather events, these regions have witnessed large-scale outmigration, social cleavages and famine. Regional food crisis had been seen in South Asia and North Africa, as agricultural lands were largely exploited. It could be one of the triggering factors behind instability and violent conflicts among inhabitants.[9]

There are other climate-induced reasons behind crop failure and food crises. The degradation and depletion of fresh water resources affect food production gravely which is more aggravated by rising sea levels. The increase in sea level could inundate substantial areas of agricultural land particularly in low-lying producing countries. In Nile delta and large portion of Bangladesh, salinization of coastal aquifers has increased, which has reduced the sources of irrigation water which have ultimately destroyed large amounts of agricultural lands and their productivity. Monsoon hits Bangladesh severely every year inundating many of its areas. The economy of the country is severely threatened by it as well.[10] The Pacific Island states are also the hardest hit where salinity affects the groundwater table and estuaries.

Along with changing climate-induced crop failures, land degradation and desertification is another phenomenon that results in tremendous human misery (as was the case with Africa mentioned earlier). This alarming process of environmental degradation is partly due to climate change but in various UN reports and by conventions it is defined as a result of human-induced factors such as cultivation, overgrazing and deforestation. It is also described as a truly global phenomenon with serious economic and social implications.[11] It had been estimated that desertification affects almost more than 250 million people. People from poor countries are affected the most.[12] Around 74 per cent of the poor are in the worst condition due to land degradation globally, a report estimated.[13]

Desertification usually occurs in the world's drylands (Figure 1.1), and climate change exacerbates it. African drylands contain over 400 million people majority of them are poor living in rural areas. These areas are degraded from desertification to a large extent.[14] Drylands are not spread uniformly between poor and rich countries. Developing countries are more susceptible than the industrial ones.[15]

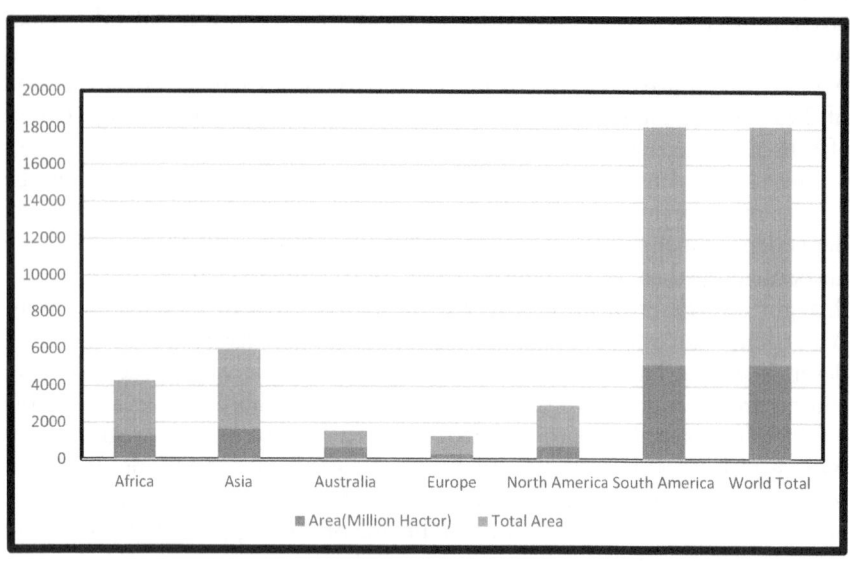

Figure 1.1 Extent and Distribution of Susceptible Drylands by Continents

Source: Prepared by author collating data from Ministry of Environment and Forest, Government of India, National Action Programme to Combat Desertification, 2002, p.10, available at http://www.unccd.int/ActionProgrammes/india-eng2001.pdf Accessed on October 20, 2014.

Desertification is impoverishing the natural potential of the ecosystems and thereby reducing the agricultural yield. In this way, it affects the food security of the people of these regions. Drylands have unscrupulously been converted into pasture and urban settlements. Escalated agricultural intensification due to growing population and scarcity of cultivable land has forced the farmers to ask for more of the land that they can yield. This rising pressure takes a toll on the land resources. In the developing world, the problem becomes more acute as with increasing need for food and declining per capita availability of croplands, people are more prone to intensified use of this already stressed resource. Therefore, countries in this region are suffering more from desertification. Environmental change has exacerbated the problem as the natural vegetation cover of drylands is usually relatively sparse and much of the soil is exposed directly to rain, sunlight and wind.[16] Climate change and rapidly changing socio-economic conditions have thus posed serious challenges to many drylands in the world. A catastrophic drought from 1969 to 1973 ravaged the grazing land vegetation and livestock in the Sahel region of Africa. It gave birth to many refugees from the devastated land as well.[17]

Sometimes food insecurity due to declining yield and land degradation may lead to direct conflicts among communities there. Though climate change induced problems are not the sole generators of these conflicts, instances across the regions have proved that societies may disintegrate due to the existence of complex nexus between environment and political and economic circumstances. The Darfur crisis of 2003 was influenced by ecological reasons at least in part as the farmers and nomads came into battle due to the nonavailability of land and water scarcity. Though the immediate cause of the conflict was a regional rebellion, the UNEP study reported that increasing desertification and lack of rain had prepared the ground for the crisis even before 2003.[18]

Like the Darfur crisis, pressure on land and water was a factor behind Rwandan Genocide too. Some studies have shown that scarcity of arable land and food crises are the driving forces behind the genocide in Rwanda in 1994 leading to a nationwide crisis. Soil degradation, faulty land distribution and continuous population growth in Rwanda resulted in nationwide environmental crises. Here, conflict over control of scarce land escalates violence.[19]

The consequences of climate change are also potential sources of revolution that have shaken the Arab world, popularly known as the Arab spring.[20] It has been argued that in the distinct chain of events that have led to the revolution climate change played a significant role. Global food price hike was a proximate factor behind the unrest which was the result of extreme global weather during the period 2010–2011. The adverse climatic conditions affected the harvest hugely.[21] Changes in food production directly affected the food supply and resulted in fluctuations in food prices. These events caused havoc to food market too. There were instances of ban on food export from Russia and Ukraine for which the Middle East and the North African regions were suffering. Food price hike though was not a principal catalyst in igniting unrest in these regions but was definitely an aggravating factor in the turmoil. Increasing food price and unemployment contributed largely to the Algerian riot in 2011.[22] Similarly, though protests in Egypt were mainly guided by the aim to topple down President Mubarak's regime, at that time the country was also suffering from food inflation that resulted in public distress too. Thus, climate change is insufficient on its own to trigger such clashes in the Arab world. But it is definitely a threat multiplier that exacerbates the already existing dissatisfaction of people due to the non-fulfillment of demands for justice, democracy, equality, political freedom and so on.

Clash of interest may also arise as a result of competing demands of different producer groups. At present, a large portion of world's energy is supplied by fossil fuel – a process heightening the risk of rapid global warming. As a result, alternative options like demand for bioenergy may rise. This would stimulate more energy crops production using many croplands for biomass energy crop yields. The same issue comes into play between different foods producing groups. Thus, competition for fertile land may intensify which is quite evident in Africa where clashes were common between nomadic and settled farmers.[23] Food production has also been hampered as a result of the utilization of cereal production for the feeding of livestock. Moreover, economic stress often stimulates the government to support cash crops production for export at the expense of subsistence agriculture. Thus, the question of land utilization brings the potential of conflict and violence between and among different producer groups.

In this way, environmental change puts at risk the very basic and universal need of humanity to get access to sufficient food. As a result, ecological constraints to agricultural production as well as the economic, social and political dimension of food scarcity may escalate the conflictual situation in some regions of the world. In such a situation if societies are already destabilized due to underemployment, ethnic conflicts and many more, the probabilities of environmental degradation leading to decreasing yields and subsequently food crises become rampant. If the state conducts its fundamental function that is guaranteeing the protection of citizen's rights to proper resource utilization as well as his or her accessibility to food and other resources well, then the risk of diminishing security due to climate change induced food crisis and land loss can be controlled to some extent.

1.2. Water Scarcity

Climate change has significant implications on water resources and their availability and quality too. Water creates problem when it is in excessive amount, when it is scarce and when it is polluted. Climate change aggravates every problem. Global warming leads to increasing temperature which in turn results in increasing evaporation rates, changing precipitation patterns affecting runoff and water quality. Rising temperature leads to increased melting of glaciers as well. The resulting increased water flows from melting glaciers are beneficial for the time being. But this further heightens the risk of flood in some regions and leads to drought events more frequent and intense in other regions. Adding to these pressures the world's growing population with their increasing demand for water resources, changing patterns of water utilities and above all water pollution exacerbate the problem of water scarcity. As a result, gaining access to clean and fresh water is increasingly difficult for many people around the world. However, environmentally induced changes in water availability affect different regions differently as the ability and the avenues required to respond to this challenge differ. The problem is further escalated as the awfully limited adaptation measures and poor water management system may harbour potential for conflicts over water, destabilization among communities that ultimately results in migration. Table 1.1 shows how climate change impacts water resources in various regions.

Table 1.1 Impacts of Climate Change on Water Resources

Regions	Effects
Africa	• Water shortage in many countries
Asia	• Scarcity of fresh water in Central, South, East and Southeast Asia • Increasing flood due to glacial melting
Latin America	• Millions would suffer water stress in coming decades and glacial retreat may result in loss of water flow
SIDS	• Sea level rise would affect the water level and salinization would increase

Source: Data collated from UNFCCC, *Climate Change: Impacts, Adaptation and Vulnerabilities in Developing Countries*, pp.18–29. Available at http://unfccc.int/resource/docs/publications/impacts.pdf Accessed on June 1, 2014.

Water scarcity directly threats food production as rain-fed production and irrigated farming consider water as the key factor for food security. With significant regional differences, agriculture accounts for 70 per cent of the world's easily accessible water available from lakes and rivers. As a result, it has to compete with urban areas and other water use sectors like industrial use and domestic household use for water supply. Therefore, dry areas in different parts of the world like sub-Saharan Africa and South Europe with heightened risks of water shortages, droughts and desertification suffer from crop failure. The surge in water demand for population density and migration pressure has further aggravated the risk of water scarcity. Thus, climate change may spur conflict among local communities and between nations over the sharing of this scarce resource.

Environmental change affects irrigated farming directly. Decreasing precipitation and glacial retreat have created havoc to this cropping system. Irrigated farming demands bulk of water in many countries like China, India, Mexico and South Africa which are suffering from acute water scarcity. Such scarcity affects the per capita food production. The irrigated farmlands in Kyrgyzstan, Tajikistan, Turkmenistan and Uzbekistan might also be suffering due to rising temperature induced glacier loss.[24] In such a situation, the Central Asian region already plagued with ethnic disputes and separatist movements become more vulnerable to weather variability as struggles over land and water resources occur on a regular basis here. Climate-induced glacial retreat and resulting water crisis have adverse impact on Peru and its capital Lima. They are potential hotspots for climate-induced water stress as they depend on glaciers for their water supply. The hydroelectric power dependent Peru's energy supply may also be threatened.[25]

In arid and semi-arid regions, water is already short in supply and most of the developing countries where predominate are suffering the most. In these regions, groundwater is significant in meeting irrigation and domestic demands. But sometimes unsustainable extraction of groundwater and lack of adequate planning and governance have aggravated the problem. Of particular importance therefore is the groundwater depletion in key food-producing areas such as the Mediterranean, Middle East, Punjab and the North China Plain.[26] Another cause of depletion of fresh water resources and salinization of coastal aquifers is the sea level rise. While soils are being degraded by the latter, extreme use of fresh water resources is decreasing the groundwater table lowering drastically agricultural production. The Small Island Developing States (SIDS) are suffering from this problem.

Rivers, both national and transboundary which are also potential sources of fresh water resources, are suffering from climate change. The amount of runoff and the rate of depletion of water within river catchments and corridors are greatly impacted by climate change. Glacial melt rivers are also affected. Snowmelt river the Rio Grande in Southwestern USA is highly vulnerable. Melting of glaciers and its retreat affect the rivers like Ganges, Brahmaputra, Salween and Mekong.[27] Glacier retreat also affects the downstream countries that are suffering from this. The problem becomes more intractable when water diversions and hydro-engineering by upstream nations take place.

Competition over the access to scarce water resource is evident in Nile River basin between Egypt and other riparian states. In this basin, the upstream states have challenged the hydro-hegemony of Egypt, thereby demanding equal use of the water. The signing of 2010 Comprehensive Framework Agreement unilaterally by the upstream states sans the approval of downstream states is the testimony of the existing current clash of interest in the region.[28] As climate change brings changes in temperature, precipitation and evaporation resulting in an increase in frequency of floods and droughts along with rising water demands due to population growth, the problem of water sharing becomes more acute. Furthermore, rising sea levels and resultant saltwater intrusion in agricultural lands rendering them unusable create food scarcity in the riparian countries and induce migration. For instance, during the world food crisis of 2008, Egypt was among the countries facing riots since people were unable to afford basic food supplies[29] and migration to cities had been induced.

Water sharing problem is also acute in the Jordan River basin. Here, water presents an environmental source of conflict, as climate change driven reduced water supply creates clash of interests among the countries like Israel, Jordan, Lebanon, Syria and Palestinians.[30] Although, in various peace processes, the hydrological issue has been taken into consideration, population growth, overexploitation, higher rate of consumption as well as arid climate and gap between the demand and supply of water have augmented the magnitude and intensity of the problem, undermining the condition for human security. Thus, water wars are most likely to occur as this issue is intertwined with the regions' deeply rooted conflicts. Having a transboundary nature, the rivers Tigris–Euphrates also create conflictual situation between the riparian states like Turkey, Syria and Iraq. The dispute between Turkey and its downstream neighbours had been erupted by a series of irrigation and dam building projects on the Euphrates by the former. It was claimed that these would have significant negative implications for Iraq's water supplies. However, the problem was further aggravated by territorial battles between Turkey and Syria where cutting off water flow of the Euphrates in Syria by Turkey would result in environmental degradation. In Central Asia, the Aral Sea basin in which Amu Dariya and Syr Dariya rivers situated also witnessed great political tensions over the sharing and utilization of water by downstream countries like Uzbekistan, Kazakhstan and Turkmenistan and upstream states like Kyrgyzstan and Tajikistan. The differences in the usage of water between them have spurred conflict affecting both the environmental and national security interests of the riparian countries.[31]

Apart from clashes ignited by the issue of cross-border water sharing, river diversion projects and dam building leading to forced resettlement of population may also act as drivers of conflict escalations. Such tensions and violence over water use rights have already erupted between India and her neighbours. In fact, China, India and Pakistan are all dependent on shared water supplies originating in the Tibetan Plateau. Since the rising water requirements of these countries are coincided with the shrinkage of these resources, their efforts to tap them ultimately foster competition and interstate conflict in the region. River diversion projects may also escalate conflictual situation in the river basin of the Mekong which is shared by Laos, Thailand, Cambodia and Vietnam.[32]

However, there are numerous evidence of cooperation among riparian states over the usage of water resource as well. Dam building for hydroelectric power generation is one of them. But due to governments' inability to maintain transparent hydrological water sharing across boundaries, the conditions for cooperative solution may decline. Another way of water pollution and resultant scarcity occurs when water resources have been targeted for political purposes. Instance of such event can be traced to the crisis of Darfur where water poisoning has been used as a military tactic in the genocidal conflict. Similarly, at Peruveli in Trincomalee district, the Srilankan military and homeguards attacked a refugee camp and dumped bodies into local wells to minimize the availability of fresh water. However, sometimes the situations are not so grim. A case of Andaman Island can be cited here where the Japanese built wells for supplying water to its troops during World War II, are restored by the residents of the city to combat water scarcity. In Syria, residents have similarly used the Roman-built water tunnel to keep their villages alive.[33]

Thus, the vagaries of anthropogenic environmental degradation may lead to tension and violence between and among states. Water scarcity, desertification and declining food production all are results of adverse climatic conditions. Paucity of usable land and water resources, adverse climatic condition like drought and floods famine may lead to movements of people to less affected region which creates pressure on the inhabitants of that region. Ultimately, the additional pressure on the resource base of a particular region due to climate-induced migration and negative impact of water scarcity on agricultural and forestry yields and on hydropower may add to the existing problems like unemployment and economic vulnerability political instability which may prove to be detrimental for nations' national security.

1.3. Deforestation

Forests provide significant ecological and social goods and services and help maintain the fertility of lands and alleviate the conditions of desertification. Besides providing wood and other products, forests are one of the most significant repositories of terrestrial biodiversity that has important value as a source of food, medicine fodder and raw material, and it contains significant amounts of sequestered carbon. However, the potential for forests to become even greater sources of carbon emission is massive. The fact is that the trees on our planet are being depleted at a very fast rate due to natural reason like forest fires or by illegal felling of trees and agricultural expansions. This process is called deforestation.

Deforestation is continuing at an alarming rate in today's world leading to massive loss of species and biodiversity. It plays a major role in affecting the carbon stock of the plants. Further, forest may become drier, leading to severe forest fires as well as could turn into grassland, steppe or desert. The biodiversity loss has emerged as a byproduct of it, and such desertification reduces the process of carbon sequestration which speeds up climate change. The most vulnerable forest ecosystem includes tropical forests, dry forests, boreal forests, cloud forests and mangroves.[34] The Amazon Rainforest, Russia and Canada's boreal forest, Indonesia's peat swamp forest and the Congo basin are the most affected ones due to deforestation. The problem has become more acute as the dying and degraded forests release the carbon they store into the atmosphere as well as affect agricultural productivity, the livelihoods of Indigenous population living within or near forest and forest-based economic activities.

Among the diverse causes of deforestation, man-made and faulty governmental policy induced degradation are the most significant. This is evident in the following cases of deforestation. The Amazon Rainforest is the largest contiguous tropical forest. Deforestation results here from shifting cultivation by subsistence farmers, commercial logging, cattle ranching, road construction and permanent agriculture and colonization programmes. The problem has further escalated when climate change makes it drier leading to more forest fires. Together with human-induced forest destruction, such phenomenon has the potential to convert the Amazon forest into savannah or a semi-desert state. It is estimated that from August 2003 to August 2004, deforestation losses amount to about 23000 km.[35] In parts of Africa, logging trails accelerate agriculture and shifting cultivation. Other causes are commercial logging and particularly settlement along logging roads. Recently, in 2019, a devastating forest fire was raging in the Amazon Rainforest area. It caused an international crisis, and the catastrophe was man-made. About 9000 sq. km was engulfed by the illegal burning of felled trees by farmers and ranchers in the region. The farmers and ranchers illegally burnt the felled trees and that caused such havoc.

The timber industry is the major cause of deforestation in some parts of Central Africa and Southeast Asia, as due to limited avenues for export; countries of these regions encourage unsustainable forest exploitation to augment greatly needed export earnings.[36] It is necessary to point that the loss of forest cover having greatest global impact is amplified by the Southeast Asian region alone. Although only 5 per cent of the worlds' forests are in Southeast Asia, 25 per cent of global deforestation is taking place because of Indonesia's huge emission level.[37] The NGO, Forest Watch Indonesia, has estimated that Indonesia with the largest tropical forests outside of Brazil is losing more than 14,000 square miles a year of forests, nearly the size of Switzerland.[38]

The Peat Swamp forest areas in Central Kalimantan which are important ecosystems in Indonesia provide many environmental goods and services both regionally and internationally. However, the natural resources here are overexploited to the point of destruction and unequally accessed and distributed by the authorities. As a result, the marginalized stakeholders do not receive adequate share of resources, and the resultant discontents ignite violence against more affluent ones. For instance, Soeharto's Basic Forestry law pushed the forest-dependent Dayak Indigenous communities into the interior by logging and plantation which spur discontent among them.[39] The adverse effect of Indonesia's uncontrolled forest fire which is the principal source of haze is another transboundary environmental assault in Southeast Asia. In 2006, the region witnessed the worse haze since 1997. The causes of the fires are both natural and man-made. Other countries of the region like Thailand, The Philippines and Vietnam have essentially lost their productive forests by logging indiscriminately. Laos, caught between timber starved Thailand and Vietnam, suffered from illegal logging of forest to supply its neighbours. It is also estimated that over a million tonnes a year of illegally logged Myanmar's timber is delivered to China creating pressure on ecosystem of Myanmar.[40] Thus, insufficiently managed natural resources, the activities by illegal loggers and plantation owners contributing to environmental destruction, negatively impact local communities, the country in particular and the global community as a whole.

Degraded forests are also found in the Russian Federation due to Industrial pollution. Chernobyl disaster also affected large area of forests in this federation. The Boreal forest is also the worst affected area and has become matter of grave concerns.

Tropical mangroves around the world are highly valuable ecosystems. Its utility to the coastal population is immense as it helps in protecting the coast from natural disasters, and people are engaged in various livelihood-related activities in these forests.[41] However, mangrove ecosystems are severely affected by climate change and human activities. Sea level rise is perhaps the greatest threat to mangroves in tropical and many subtropical areas of the Caribbean, Latin America and Africa. Along with climate change, many changes in the coastal regions due to economic development, land use practices and demographic shifts have affected these areas too. The mangroves in many South and Southeast Asian countries like Bangladesh, India, Srilanka, Myanmar, Thailand and Indonesia are affected by human activities like change in agricultural practice, engagement in aquaculture and industrial logging. Central America and Northern Australia are highly affected by the industrial logging induced mangrove loss. Africa is also suffering from mangrove loss due to expanding agriculture, rapid urbanization in the coastal area and oil pollution.[42] Biodiversity loss has been resulted from mangrove loss too affecting the coastal ecosystem. In such a situation, the coastal areas have become more fragile. Though climate change is not the sole driver of this kind of deforestation, it definitely adversely affects the ecological balance and speeds up climate catastrophes resulting in tremendous human miseries.

In different regions, deforestation has affected the climate as well as the lives and livelihoods of people. As there is a wide range of competing uses of forest by diverse groups, like some depend heavily on forest for their subsistence while others may have a higher level of industrial or artisan skills and access to market, deforestation may lead to violent conflict between those who exploit and those who conserve. Besides, forest degradation and deforestation are potential enough to deprive the Indigenous population of their natural habitat and tenure rights in forest areas culminating in social disruption. These groups are often excluded either deliberately or through fragile governance system of the state from adequate engagement in the commercial use of forests. The problem has further been deepened by cross-border illegal trading of timber leading to scarcity of fuel wood in the developing countries, where alternative energy sources are too expensive to afford. Further, the Indigenous people may themselves affect the forest ecosystem by altering the existing forest landscape in ways that may not in congruence with the interest of global stakeholders. Moreover, the absence of sustainable forest management may have negative impact on forest-based export earnings which in turn lowers government revenue and limits the options for a diversified economy. Instead of being a means to alleviate poverty, destruction and mismanagement of forest increase poverty. All of these may sow the seeds of discontent among the marginalized local communities which may take the form of violence and sometimes mass migration to cities. The failure to manage forests sustainably, to harness the potential of it, to reduce poverty and to ensure that women, poor and all other deprived groups are actively participating in the formulation of forest policies and programmes have tremendous environmental consequences globally and locally and over the impoverished population across the world.

1.4. Problems Pertaining to Energy Resources

The demand for energy resources is surging in the world today. The fossil fuels, mainly oil, coal and natural gas, are today meeting most of the global energy needs. Combusting fossil fuels have serious implications for the environment, and they are the sources of GHGs causing global warming and climate change. For a growing economy, a reliable, affordable and sustainable supply of energy is required which ensures the energy security of that economy. So countries are facing the dilemma of economic growth based on sustainable supply of energy resources without harming the environment. It is true that too much use of fossil fuels harms the environmental security of the world, but it affects different countries differently. Against such a backdrop in today's world, each and every government requires a coherent policy to address both the energy and environmental security issues.

Oil, gas and coal are of fossil origin and are playing crucial role in yielding energy and dominating future energy markets. Nearly, 60 per cent of GHG emissions of the world come from burning fossil fuels. The problem has been augmented as the reserves of these resources are unevenly distributed across the world. Therefore, countries even the developed ones are required to depend on import of some of them according to their respective demands.

The varied degree of import or export of energy resources determines each country's perception of energy security. For instance, while Russia and Saudi Arabia are almost self-sufficient with regard to energy, relying on their vast domestic resources of oil and natural gas, the United States though is self-sufficient in supplying energy for the electricity sector depends on oil import for the transportation sector. The Europeans also depend on imported oil but their energy security is mainly influenced by their dependence on Russian natural gas.[43] With industrialization and urbanization, transportation has increased in the developing world and as their lifestyle changes, the non-OECD countries consume more global oil. But they have limited resource endowment that results in their dependence on foreign energy resources for both electricity generation and transport sector. So the global community is facing the dual challenges of energy security and global warming.

Despite the associated threats, use of fossil fuel is not declining in non-OECD countries. Subsidized oil products and coal are responsible for that. Gas is cleaner in comparison to coal. Liquified natural gas (LNG) which is today about 8 per cent of all internationally traded gas helps in improving energy security by fostering diversification.[44] The inefficient use of energy resources in developing countries and rapid industrialization in the emerging economies that have increased the energy intensity of these economies have contributed towards rising trends of GHG emissions from fossil fuels.

Most of the oil reserves are traced in the Middle East, Canada and Russia. The ability of the oil market to meet the rising demand will largely depend on the availability of reserves.[45] Regarding oil market, OPEC retains inordinate power. Countries which are mainly oil consumers have to depend on imports that might have economic and political consequences. But as oil demand is likely to rise and it is concentrated in politically unstable region of the earth like Persian Gulf, any crisis disrupting supply may have tremendous impact on global economy. Energy security problem with reference to oil is therefore getting worse. Its uneven and declining reserves across the world, high finding cost, and demand and supply imbalance have instigated the search for alternative resources like natural gas and coal which has its own shortcomings with reference to both energy security and environmental concern.

Natural gas emits fewer greenhouse gases than oil, and there is a large-scale demand for gas to produce electricity despite the fact it is expected to run out in a little over 50 years.[46] Huge finds have been discovered in many parts of the world, and the largest or 'elephant' fields are in the Middle East (Iran and Qatar) and Russia. Gas, an environment-friendly alternative to oil, has given its producers a monopoly over the new resource. The gas reserve in Qatar is the third largest in the world following Russia and Iran.[47] It is attractive in environmental terms, and its consumption is growing as new discoveries are taking place. But it has its own limitations. Compared to oil, it is expensive to transport over long distances and it requires long-term commitments between producers and consumers. As the gas supply is monopolized today by limited suppliers, its price goes up. Political instability in gas-producing countries and their strained relations with transit countries have resulted in supply disruption. For instance, the Russia–Ukraine Gas war of 2006 has resulted in supply disruption in Europe as Ukraine is the only transit country between them. Europe has faced energy insecurity with respect to natural gas several times. Therefore, the Gas era has witnessed crucial interruptions of its vital supplies, and sometimes due to market dysfunction a large amount of gas in the world is flared off. LNG is considered a better option for transport and storage of natural gas. Qatar accounts for the largest export of LNG whose transport cost is also very high.[48] Asian countries like China and India need the supply of LNG in bulk, but the cost issue may act as an impediment in using largely this environment-friendly resource.

The use of compressed natural gas (CNG) in transportation sector is another option that addresses both the energy security and climate change issue. The US Environmental Protection Agency has estimated that substituting CNG for gasoline would lower carbon dioxide emissions by about 25 per cent. But along with cost issue there is also problem of leakage that has turned use of CNG in vehicles critical. When natural gas leaks, it enters the atmosphere as methane which is a GHG. Industrial operations can be designed to reduce leaks to a negligible level; filling stations operated by average individuals, however, risk having small leaks that would overwhelm any emission reductions from using natural gas in vehicles.[49]

However, since both the reserves of oil and gas are limited and there is spectacular increase in the price of oil, there is a tendency to rely on coal which is more widely available and is abundant and relatively cheap, to meet the accelerating energy demand of emerging economies and developing countries. Therefore, there exists minimum energy security problem as countries are not largely dependent on imported coal. This temptation of relying on coal has a tremendous detrimental environmental effect as it is estimated that coal usage discharges 9,000 million tonnes of carbon into the atmosphere every year.[50] So the fundamental impediments to further coal use in electricity generation and to fulfil other energy demands are environmental considerations. In such a situation, what is urgently needed is the progress on clean-burning technologies; otherwise, countries may face increasing restrictions and disincentives on its use. However, in 2020, Coal use experienced large decline in the United States and the EU. The growth of renewables this year at the expense of decline in coal use is noteworthy in this regard. But renewables have their own shortcomings.

Nuclear energy is cheap, reliable and CO_2 friendly. Nearly, about 16 per cent of world's electricity production comes from it. Such power plants are mostly concentrated in the United States, East Asia and Europe.[51] But the spread of nuclear power may have serious environmental and security risks. The accidents in Chernobyl and Fukushima nuclear power stations had serious consequences for the environment and human lives. There are also problems with the storage of nuclear waste and use of nuclear power for building WMDS and the fear of nuclear attack on the power stations. Moreover, International Atomic Energy Agency estimates that economically recoverable identified, assured and inferred uranium reserves are not sufficient with respect to demand. These weaknesses have impeded the use of nuclear power as a viable option for both energy security and environmental protection. Other renewable resources like water, wind, solar and biomass are also sustainable, available and climate friendly, and they can be used to confront the twin challenges of energy security and global warming. The proliferation of these resources will help in reducing net GHG emissions, while their domestic abundance may reduce dependence on foreign energy resources. The use of renewables like solar and wind power by a country does not limit others to use the same within their territorial borders.

Therefore, the increased use of these resources is envisaged for enhancing both the climate and energy security. But there are also some challenges and geopolitical problems. Hydropowe, which is considered as cleaner source of energy if compared to other sources, has its own limitations. Climate change induced dwindling water flow of the rivers is an impediment in generating hydel power. On the other hand, dam building on an upstream river may bring changes in the riverscape affecting the flooding pattern as well as the lives and livelihood of people living downstream. It may give birth to political animosity between the upstream and downstream users within a country and between the countries if the river basin is shared. Solar and wind energy resources are considered environmentally benign. Due to their abundance, they can enhance energy security. But in the absence of adequate electric storage capacity or backup power generators, wind and solar, with the exception of solar thermal plants, are unable to provide base-load electricity and are prone to interrupted supply.[52]

The growth of renewables has witnessed a remarkable spike in 2020. The energy market has been dramatically impacted due to COVID-19 pandemic this year. Wind power, solar power and hydro-electricity generation have increased during this period despite the drop in total energy demand, especially in the United States, India and Russia. Most of the growth in renewables was found in China, the United States and Europe.[53]

So the nexus between energy resources and climate change is vital. The formulation of energy policy is a domestic venture, but as it is related to energy markets and environmental problems, there may exist clash of national interests between and among states. Sometimes the energy policies of the producing countries may affect the energy security of the consumers. The mood and fancy of the transit states are also affecting the energy security of the consumers. Though there are today many producers of energy resources, they barely meet the surging energy demand. Therefore, there is fierce competition among the major oil and gas-consuming nations. Climate change has turned this battle more complicated for both the producers and the consumers. Production and the use of fossil fuel based energy is the main source of GHGs, which is a proven fact today. But the process of mitigation and adaptation may harm energy security. As the producers' economic well-being is contingent upon incomes from oil and gas export, climate change related policies that put barriers to the usage of these resources may cause harm to their economy. Sometimes use of renewables may enhance both the energy and ecological security of the developing countries. But many of these measures are more expensive than the fossil fuel alternatives. In such a situation, developing countries who are mostly dependent on commercially traded fuels from the OECD countries may opt for cheaper fossil fuels like available coal and inefficient use of non-commercial resources like biomass, thereby affecting the environment and causing health hazards. To increase the cost of coal as a means to prevent its use may deny the rights of the poor multitude to basic electricity and fuel security as well. They are thus caught up by the energy poverty trap that diminishes their energy security too. Additionally, uneven distribution of gas resources and the supply-side geopolitical and commercial complicacies may minimize its use as a mitigation measure for replacing coal. Thus, energy efficiency, energy conservation, the efforts to promote cleaner energy which help in enhancing energy security and emission reduction, sometimes certain climate policy objectives, may stand in contradiction with some countries' understanding of energy policy objectives. Therefore, such policies could be detrimental for address-ing energy security as it becomes more and more complex.

2. Extreme Climate Conditions

Climate change is a slow onset process that gradually affects the environ-mental and weather patterns of countries. Due to its adverse impacts, sea level is rising in some places of the world which is a silent process but has severe consequences for coastal and low-lying regions of the world. In con-trast, extreme weather events like cyclones and storms arrive quite rapidly with devastating effects on lives and livelihoods of people.

2.1. Sea Level Rise (SLR)

It is projected by a report that within 2100, due to increase in global mean surface temperature of up to 6.4 degree centigrade, almost 59 cm SLR would result.[54] Such a rise could cause considerable alteration of some coastal envi-ronments which may lead to higher risk of erosion, increasing saltwater intrusion and more frequent and severe flooding of low-lying coastal areas as well as serious disruption of existing biophysical systems and to human habitation and livelihoods.

The low-lying deltaic regions of Bangladesh are considered to be among the vulnerable areas at greatest risk from the adverse climatic effects includ-ing sea level rise, tropical cyclones and floods. In particular, sea level rise may turn the coastal areas more saline rendering them less fertile for agri-cultural production. In India, the Bay of Bengal sector comprising states like West Bengal and Orissa, in 2050, nearly 1 million people, would be at risk as estimated in a report by Norman Myers. Bangladesh would be affected more due to SLR during the same period.[55] Egypt is also suscepti-ble to this crisis as intrusion of saltwater created havoc on its agricultural production.[56] Sea level rise is also a threat to the economy and environment of the Mekong Delta region having adverse impact on the rice production in Vietnam. The low-lying region of Mekong Delta and red river where the most productive lands are concentrated were also affected. The populations of these areas are severely impacted as valuable arable land, coastal fisheries might disappear, Port facilities may be hampered, coastal industries may be lost, and transportation may be disrupted.[57]

In many Asian cities, industrial activities are located in the coastal sec-tor. A large amount of China's GDP is generated in the coastal area. So SLR affected these regions to a great extent.[58] The situation is more grave in the Island states like Kiribati, Maldives, Marshall islands and many such Pacific and Caribbean atoll states. They are suffering from existential threat as many of them are on the brink of complete submergence due to rising SLR.[59] Their dependence on coastal ecosystem for economic purpose

and the tourism industry specifically is hampered. These are small economies, and their reliance upon few sectors, on imported goods, is at risk due
to SLR.

With the projected rate of sea level rise Pacific Island states like Kiribati and Tuvalu, exclusive economic zone might reduce as a result of inundation as well as a loss of seas decreasing the area of exclusive rights for
marine resources and reducing potential income. Especially at risk will be
those areas where the coastal populations face displacement due to complete submergence of entire territory. As Tokelau and the Marshall Islands
are afflicted by New Zealand and the United States, respectively, there is
a hope that the displaced may resettle there. However, in case of Kiribati
and Tuvalu, as such international connections are absent, the environmental refugees from these islands are to relocate elsewhere to seek sanctuary.[60]
Specifically, Tuvalu in the Western Pacific has endured lower level flooding,
penetration of saltwater and increased coastal erosion as a result of rise in
sea level. Although the population is only about 15,000, if the situation
worsens further, the entire population has to leave their original place.[61] No
decision has yet been reached on the plight of statelessness of the Tuvaluans
by the New Zealand government, and there exists little or no protection for
the stateless people. Host countries quite naturally suffer from xenophobia as such complete extinction may proliferate socio-economic upheavals
and may amplify cultural and ethnic problems and extensive political chaos
there. In some of the Pacific states where highlands are in existence as are
the cases in the Solomon Islands, Vanuatu and some others, there may be
internal displacements. But such internal displacements have their own limitations. Land rights which are communal may affect such migration in the
Pacific.[62] Rising sea levels also affect Maldives consisting of nearly 1200
islands. The Government of Maldives refers to the country as an endangered nation as the effects of global warming could create a volatile situation within the country. It would be catastrophic during the storm surges for
these STDS even due to a 1-m rise in sea level.[63] Climate change will affect
the economy of Maldives gravely too. Their heavy reliance, like other Small
Island Developing States on oceans and fisheries for both food security and
livelihoods, is also severely threatened by rising sea levels.[64]

For these countries, failure in the efforts to adapt to changes in climate is
detrimental for these island states to a larger extent. If they fail to thrive in
a changing climatic condition like sea level rise and storm surges, it would
not only alter their physical structures but would also affect their economies, uproot a large segment of population, augment resource scarcity and
thereby ultimately threaten the national as well as international security
interests of these fragile countries. So sea level rise not only affects the food
production and fresh water availability of the coastal areas across the world
but would also trigger enormous economic and social dislocations as well
as induce large-scale migration and disruptions to human habitation and
livelihood.

2.2. Increased Intensity of Natural Disasters

Cyclones, windstorms, high level of precipitation and associated floods, tsunamis and earthquakes are loud natural disasters that have claimed human lives and caused destruction to property on a large scale across the globe. The intensity of these disasters has been increased due to climate change as well as by human actions that have augmented the fragility of already ecologically vulnerable regions. Natural disasters affect both the developed and developing worlds but due to the low adaptive capacity and lack of disaster management facilities in the latter, they generate more economic loss and fatalities there.

In the Bay of Bengal, tropical cyclones and heavy storms are very frequent. Climate change has increased its frequency and intensity. Bangladesh is the worst victim of such disasters. During the pre-monsoon (April–May) or post-monsoon (October–November) seasons, cyclones frequently hit the coastal regions of Bangladesh. In the past 50 years, most of the global storm surges are recorded in the Bangladesh. The country is experiencing every year huge casualties for that.[65] Cyclone Sidr was one of such cyclones. It hits Bangladesh coast in November 2007, with wind and tidal surge and killed at least 3,000 people, levelled homes and forced the evacuation of 1.5 million people.[66] Cyclones Vardah, Foni, Umphan and Yash have also caused tremendous economic loss and loss of lives in the Eastern India and Bangladesh in 2016, 2019, 2020 and 2021, respectively.

In recent past, there were other instances of deadly cyclones all over the world. At the end of August 2005, a powerful hurricane named Katrina made landfall along the Gulf Coast of the United States, killing approximately 2,000 people and revealing significant inadequacies in the federal, state and local preparation for and response to these events. In May 2008, another major Bay of Bengal storm, namely, Cyclone Nargis, struck Burma. The Burmese government's official death toll was approximately 22,000, but it had been said that the actual death toll might have exceeded 100,000.[67] In February 2011, Cyclone Yasi hit Queensland, Australia. The cyclone was 500 km wide with an eye of 100 km in diameter and 285 km/hr wind speeds.[68] In 2013, there were several devastating storms affecting many countries with huge economic loss and loss of lives. Due to Typhoon Haiyan, the Philippines experienced nearly $ 13 billion in economic loss, and about 6000 people lost their lives. Cyclone Phailin affected India this year also leading to heavy economic loss. In terms of extreme weather events, 2013 will most likely be remembered by Typhoon Haiyan, which struck the Philippines in November 2013, inflicting over in economic loss. In the same year, tropical storm Manuel hit the West Coast of Mexico, while Hurricane Ingrid affected the East Coast. The country witnessed nearly $ 6 billion in economic damage, and massive landslides in the affected region were also huge.[69] The year 2017 was marked as the costliest year in terms of the climate disasters across the globe. Hurricane Maria hit Puerto Rico this year and caused severe damage to this unincorporated American territory. In the year 2019, there were instances of such climate catastrophes too. An intense tropical Cyclone Idai caused severe damage and humanitarian crisis in Mozambique, Zimbabwe, and Malawi in March 2019. Kenneth was another cyclone which affected Mozambique hugely just some weeks after Idai hit the place. It was considered the strongest cyclone ever recorded on the African continent. Typhoon Texai and Hagibis were also deadly that created havoc in Japan in 2019. Typhoon Hagibis was considered the most powerful typhoon in Japan in more than 60 years. Similarly, Typhoon Lekima hit China hardest in the same year. The Bahamas was also struck by Hurricane Dorian in the same year, and it was of high magnitude.[70] Sometimes such disasters have cross-border repercussions as happened in 2015 when Cyclone Roanu hit Srilanka. It caused damages to India and Bangladesh also. Thus, the year 2019 was marked by violent changes in the atmosphere associated with extreme weather phenomena leading to some of the deadliest disasters.

The frequency of tsunamis and earthquakes is also enhanced. In 2004, one of the largest magnitude earthquakes of the previous century struck at a shallow depth off the coast of Indonesia. It was followed by tsunami waves which slammed into the coast of an Indonesian island Aceh. The waves had propagated in all directions from the earthquake's epi-centre, eventually crossing the Indian Ocean. Over 250,000 people were killed in more than a dozen countries including Sri Lanka, with Indonesia experiencing the most devastating effects and the highest mortality. Sri Lanka then suffered the second highest mortality with 30,000 dead and approximately half a million homeless.[71] The Andaman and Nicobar Islands and Tamil Nadu coast of India were also affected greatly. In 2018, Indonesia witnessed devastating earthquakes and tsunami again starting from July till December. The earthquakes on the island of Lombok in West Nusa Tenggara Province of Indonesia created huge casualties. Another tsunami struck the coast of Central Sulawesi triggering landslides and soil liquefaction in several densely populated districts, burying entire villages in September 2018. In December 2018, Anak Krakatoa, a small volcano in the Sunda Strait, erupted and generated another sudden tsunami that hit the densely populated coasts of Java and Sumatra on either side of the strait. All of these three events led to death of more than 3000 people, and more than 700,000 people were injured or displaced (Asian Development Bank, 2019).

In recent periods, due to climate change both violence and frequency of torrential rain events have been increased. Heavy rainstorms often lead to flash floods in various parts of the world. Moreover, glacial retreat and rapid melting affect volume of river water which results in floods in some places while water scarcity in some other regions. India is suffering from several devastating floods in the recent past. Heavy rains flooded China, Pakistan and Colombia in 2010. In the period 2011–2012, countries like Brazil, Thailand, Japan and Great Britain witnessed several devastating floods that claimed thousands of lives and caused heavy financial losses. The Uttarakhand of India faced dangerous landslides and floods caused by torrential rain in 2013 as well. Prolonged monsoon in India has now become a common event creating heavy floods in recent years. In 2019, such torrential rains created catastrophes in 14 states of the country rendering many homeless. South Sudan and Niger have also experienced tremendous flooding during this year. In 2021, again Uttarakhand and North Bengal of India were flooded by intense monsoons leading to death and destruction in these regions of the country.

Though storms, precipitations and flooding are periodic phenomena and take place at regular intervals, climate change has escalated their intensity and frequency. It is evident in the fact that increasing number of natural climate disasters (Figure-1.2) across the globe are hitting the headlines nowadays.

All of these natural hazards have detrimental effects on economy and socio-political structure of the affected countries. An inevitable result of such events is the uprootment of people from their original habitat. Such migration requires special analysis as it has enormous security implications for a country and is a challenge for not only interstate relations but also intra-state politics.

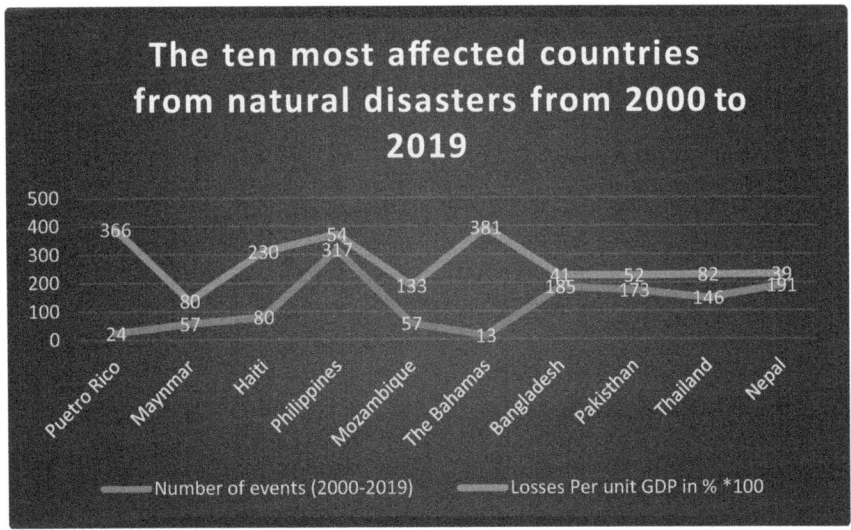

Figure 1.2 The Ten Most Affected Countries From Natural Disasters From 2000 to 2019

Source: Prepared by author collating data from David Eckstein, Vera Künzel, Laura Schäfer, *Global Climate Risk Index 2021 Who Suffers Most from Extreme Weather Events? Weather-Related Loss Events in 2019 and 2000–2019*, Briefing paper, Germanwatch: Germany, January 2021. Available at Germanwatch Accessed on January 15, 2022.

3. Environmental Change Induced Displacements

The world history has repeatedly been witnessing the phenomenon of large-scale expulsions of population, forced to flee their homes due to varied reasons ranging from political violence, ethnic and religious problems to economic chaos and disastrous natural calamitous situations. However, the environmental refugees, unlike traditional refugees, who are not recognized by the Geneva Convention of 1951[72] or the United Nations High Commission on Refugees (UNHCR), do not have the same legal standing in the international community like the conventional refugees. The international refugee regime based on 1951 Geneva Convention does not provide protection to these people who flee for environmental reasons.

Although the precise definition of climate refugees varies commentator to commentator, there is a broad consensus regarding the features of them in all interpretations. These are people who migrate because of environmental factors. Lester brown first popularized the concept in 1970s and later in 1990s, El-Hinnawi explained the term elaborately. He underscored three broad types of environmental refugees, namely, refugees who are temporarily displaced due to natural- and human-induced disaster; permanently dislocated people due to development works like construction of dams; and migrants who are uprooted for gradual environmental degradation. He added another type of environmental migration where displacement occurs due to war-ravaged environmental destruction.[73]

Due to the absence of an internationally agreed definition, the International Organisation for Migration(IOM) interprets a working definition which defines these clusters of people as follows:

> Environmental migrants are persons or groups of persons, who, for compelling reasons of sudden or progressive changes in the environment that adversely affect their lives or living conditions, are obliged to leave their habitual homes, or chose to do so, either temporarily or permanently, or who move either within their country or abroad.[74]

Along with outmigration due to environmental change, the definition by IOM also pinpoints the internal displacement of people. These are the people who are displaced due to natural disasters, developmental activities like dam and road building, urbanization, mining and so on but stay within their own borders (Figure-1.3). However, the UNHCR has acknowledged the plight of IDPs in paragraph 2 of the guiding principles on International Displacement.[75] However, such efforts have narrow applicability due to non-enforceable legal status of these IDPs.

There is also widespread debate regarding the pattern of mass mobility due to fast- and slow-onset environmental factors. During natural disasters, people have generally little choice but to move within the national boundary and sometimes after initial setbacks and receiving relief, they may return

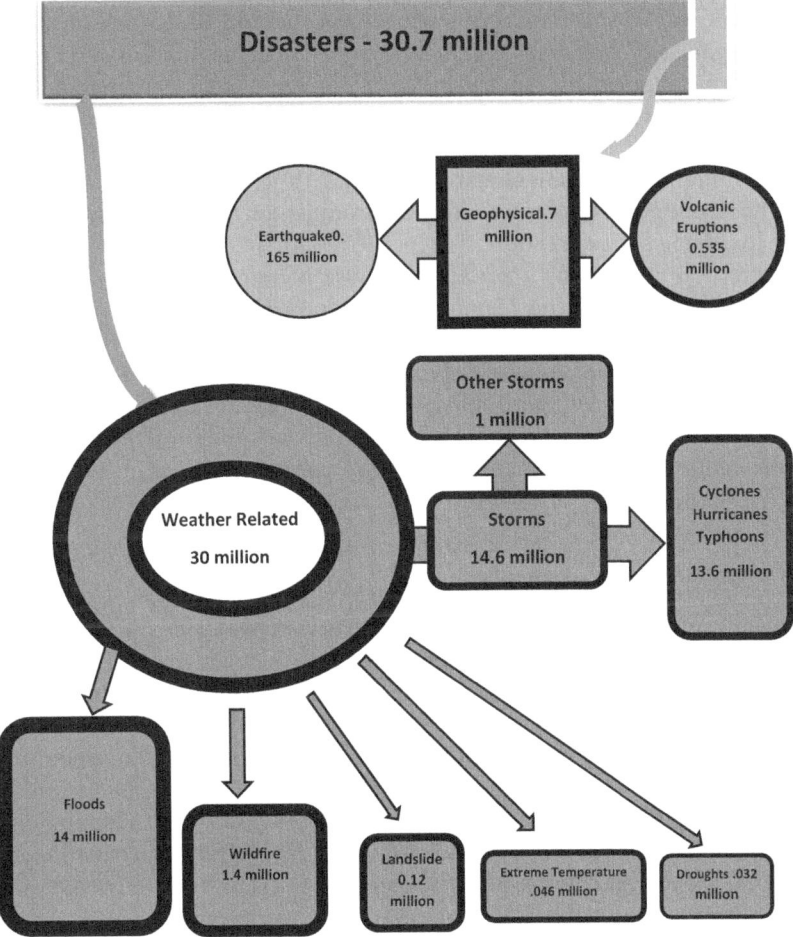

Figure 1.3 Internal Displacements by Disasters in 2020

Source: Prepared using data collated from IDMC I GRID 2021 I 2021 Global Report on Internal Displacement (internal-displacement.org)

back. Such temporary dislocated persons are named evacuees. In 2005, hurricane Katrina gave birth to nearly two million temporary homeless people on the gulf coast of the United States.[76] However, during gradual- and slow-onset climate change, movement of people is of voluntary nature. Moreover, there remains a wide grey area in the sphere of identifying forced and voluntary migration. The problem has further been augmented by the difficulty of distinguishing environmental refugees from economic refugees, as environmental stresses often intersect with economic vulnerabilities.[77]

Various reports have estimated the scale of environmental displacements. One of the popular estimations has been put forward by Norman Myers that says in 1995, the total of such refugees would be 25 million if compared against 27 million traditional refugees.[78] This estimation of 25 million had been endorsed by the 2001 World Disasters Report prepared by the Red Cross and Red Crescent Societies.[79] An estimation by the International Organisation for Migration (IOM) also calculated that in 2050, there could be 25 million to 1 billion climate refugees.[80]

The developing countries and the environmentally vulnerable geographies like low-lying states are more susceptible to climate change. Environmental degradation appears as a proximate cause of migration here due to low adaptive capacity. Of the 25 million environmental refugees in 1995, most of them are mapped in the African Sahel and Horn of Africa, while in Sub-Saharan Africa, many people were starving due to ecological stresses which resulted in the uprooting of millions of them.[81] Recent data by IDMC have projected that as of 31 December 2020, at least 7 million IDPs exist across 104 countries.[82] Environmental migration can be triggered by various factors, and they can be categorized in the following ways.

3.1. Drought-Induced Displacements

Sahelian zone is the prime locus of such displacements. Here, the accelerated frequency and intensity of droughts and floods together with large-scale demographic and livestock growth are potential enough to drive refugee flows. During 1970s and early 1980s, the region witnessed severe drought situation which culminated in both internal displacement and outmigration. The close relation between the state of environment and migration can well be conceived if one concentrates on the nomad strategy of responding to seasonal changes in environmental condition by migrating across the area. Besides, the entire region is also suffering from severe water shortage which may induce southern movement of the groups. Moreover, this area is affected gravely by land degradation due to expanding foreign exchange earning commodity production. Such encroachment on land traditionally used by pastorals gave rise to violence as the nomads fought back. Sometimes, they intruded on the land of others. Before their complete displacement, the nomads came into conflicts with small-scale farmers and sometimes became able to strike back at the state itself, as was the case in Chad, where nomads of the desert zone formed bands of revolutionaries.[83] In Mali, during the 1983–1985 droughts, distress migration albeit temporary took place. In the Faguibine area in Mali, extended dry season led to environmental migration dislocating nearly half of the population.[84]

In the drought-prone areas of Ethiopia, a 10-year (1984–1994) period study unfolds the truth that rates of outmigration were highest during the periods of the greatest drought-driven food insecurity. The country heavily suffered from displacement again while facing the strongest onset of El-Nino in 2015. It resulted in long-term drought lasted till 2019, in the southern and southeastern part of the country. In an estimate that shows up to March 2020, 381,426 people were displaced by drought in 244 sites, which equals 22 per cent of the total number of IDPs in the country.[85] In Senegal, an area called Tambacounda was affected by the soil erosion in the early part of 1990s which resulted in the migration of male population of the area. The women were left to the native area with huge economic burdens.[86]

In other regions of the world, drought-triggered migration flows are also significant. In large parts of Brazil's northeast, along with human-induced land degradation, drought conditions forced local farmers to undertake out-migration as a survival strategy too. For instance, the Sertao region witnessed internal rural-to-urban displacement of some 60 million people during the period 1970–2005 due to prolonged and continuous dry seasons.[87] Migration as a response to droughts is also evident during the Dustbowl years of the 1930s. During the dustbowl decade, 300,000 people left the Great Plain States. Many of them went to California because of climate events like multiple years of below-average rainfall and above-average temperature which coincided with the great depression resulting in a worst situation for the Oklahomans.[88] In Somalia, both drought and flooding have ravaged the country almost every year. In 2017, more than a million people were displaced due to drought. In 2020, nearly 1 million people in the country were displaced for environmental reasons. Even amidst COVID Pandemic, in the early months of 2021, many people were displaced of whom more than 30 per cent were drought induced.[89] The IOM has also monitored that in 2018–2019, large influx of migration which was mostly drought induced from the surrounding provinces took place in Herat City and Baghdis (Afghanistan).[90]

3.2. *Extreme Weather Induced Displacements*

As the intensity and frequency of disastrous events have been increased as a result of climate change, they also help in originating large numbers of climate refugees. Flood-related riverbank erosion in the delta region of Bangladesh affects nearly 1 million people every year leaving them homeless either temporarily or permanently. Regular river floods are proved to be detrimental for medium to large areas of the countries, while during extreme years, floods engulf almost 70 per cent of the countries. The years 2004 and 2007 marked such extreme years while people had to relocate either temporarily within the country or sometimes beyond border.[91] It is estimated that 1998 monsoon floods in Bangladesh inundated two-thirds of the country for 2 months leaving 21 million people homeless.[92]

Heavy rains and seasonal flooding also trigger migration across the world. Many countries in Africa are suffering from forced displacement for such reasons. The southern and eastern parts of Angola in 2009 were ravaged by floods leaving many displaced. It is estimated that in the period 2003–2008, annual flooding alone displaces more than 170,000 people in Ethiopia. Madagascar is one of the most vulnerable countries in Africa to tropical cyclones. In 2008, the country was hit thrice by cyclones. Countries in Southern Africa like Mozambique and Zimbabwe are also suffering displacement due to such negative effects of climate change.[93] Tropical cyclones accompanied by storm surges also affect Bangladesh affecting damage to the resources. The cyclones in Bay of Bengal affect the country every year leaving many homeless. Cyclone Sidr was one of them that affects the country in 2007. Recurring cyclones not only in Bangladesh but also in Myanmar are evident.[94] Cyclone Nargis uprooted many people in Myanmar in 2008.[95] In 2020, Cyclone Amphan displaced almost 5 million dislocations in Bangladesh, India, Myanmar and Bhutan that is recorded as the largest in the year.[96] The frequent cyclones in several parts of the world created a huge number of climate migrants and internally displaced persons in 2019–2020 like Cyclone Idai of 2019 created a large number of internally displaced persons in Mozambique.

Hurricanes are also responsible for displacement. The southern part of Mexico was hit by hurricanes Emily, Stan and Wilina which uprooted people from the affected places in 2005. Hurricane Katrina did the same in the United States in the same year displacing more than 1 million people.[97] Thus, harsh environments wreak havoc on human settlements disrupting livelihoods and displacing people temporarily or permanently for sheer survival.

3.3. Sea Level Rise Induced Displacements

Due to sea level rise, there are also displacements. Norman Myers' estimation is that 26 million in Bangladesh, 12 million in Egypt, 73 million in China and 20 million in India are at risk of SLR-induced displacement.[98] IPCC calculates that 45 cm rise in sea level would displace 55 million people and submerge over 10 per cent of Bangladesh.[99] It is also estimated by some experts like Hefin Jones, from Cardiff University, that in the coming 50 years more than 10 million climate refugees would be recorded in Bangladesh while China and India would produce 30 million each.[100]

Like the Gangetic plain of Bangladesh, the Nile Delta of Egypt which is the breadbasket for the country is extremely vulnerable to sea level rise. Particularly, the North Eastern coastal zone of the country is severely threatened. It has been projected that 50 cm rise in sea level by 2025 would generate more than 2 million displaced people in the delta.[101] Thus, human-induced climate change can hit the region with ferocity which may culminate in both internal migration and potential regional and international relocation.

The future of atoll states is also questioned as severe climate events might ultimately lead the inhabitants to relocate elsewhere. It has already been discussed how Tuvalu (a Pacific Island state) is affected by rising sea level and extreme weather events. Its 11,000 residents are in serious trauma of being displaced by sea level rise. Although some of them have already relocated to Fiji, New Zealand and other neighbouring islands, the remaining residents face an imminent threat due to the same. Although the agreement of Pacific Access Category, as agreed between New Zealand and Tuvalu established a specific immigration deal to enable environmental refugees to take refuge in less vulnerable region, it is limited to those aged who are middle aged, thereby representing a mere economically oriented policy to bolster New Zealand's workforce.[102] Carteret island of Papua New Guinea was the worst affected in Pacific. In 2005, Carteret Islands was evacuated by a political decision, so resettlements of the islanders were followed to the larger Bougainville Island.[103] Such events are the testimonies of the fact that climate migration is inevitable in the face of slow and rapid onset of disasters.

Along with these previously mentioned catalysts, there are some other reasons behind climate change driven displacement. Deforestation affects the Indigenous population largely as they are forced to migrate from their traditional habitat. As was the case with Indigenous population of Amazon Rainforest and many scheduled tribes of India who have suffered as deforestation and commercial use of forest land have restricted their access. Such uprootment of the traditional people and severance from their own culture force them to relocate to a place alien to them and turning them impoverished. Similarly, in Haiti, deforestation leads to soil erosion which accelerates poverty and gives rise to large-scale outmigration.[104] Apart from deforestation, land degradation also causes outmigration. In Indonesia and Indian state of Punjab, as a result of land degradation due to intensive cultivation, both in- and outmigrations took place.[105]

The melting of glaciers in mountain regions may also threaten the existence of communities living in lower valley. In the Himalayan region, glacial lake flooding may result in large-scale fatalities and displacements. Furthermore, sudden natural disasters like tsunami and earthquakes also induce large-scale uprooting of population. This is evident in the incident of 2004 tsunami due to which mass migration took place in the South and Southeast Asian region.

It has thus been proved that there exists a real and complex relationship between environmental change and migration. However, displacement whether forced or voluntary is not always just product of environmental push. There may exist a complex nexus between environmental and socio-economic factors that prompts displacement. So climate change induced migration has far reaching social, political and economic repercussions. It does not trigger conflict directly. But it can alter the ethnic composition within and between states and sometimes can intensify instability and clash of interests in situations of resource scarcity. Climate migrants may place

considerable stresses on natural resources of the destination place resulting in a competition for access to resources between insiders and aliens.
Such scarcity-related insecurities can contribute to the unproductive conflict based on ethnic inequalities. In Ethiopia, the refugee community of
Bonga camp in the Gambella region of Western Ethiopia had created pressure on the land and forest resources and gradually outnumbered the host
communities. The host communities were also alarmed at the population
growth of Uduk community. Although Bonga has not suffered any open
communal violence suffered elsewhere in the region, there were environmental resource use conflicts took place in Itang and Abodo in 2001 and
Crog in 2002. In Fugnido camp, near Bonga, 42 refugees were killed due
to ethnic clashes in November 2002 forcing 531 refugees from Fugnido to
Bonga to relocate.[106]

Apart from such ethnic clashes and resource war, a kind of xenophobia
among the inhabitants of the host country comes into prominence which
has further aggravated the problem leading to social turmoil. In developing
countries with increasing population pressure, an additional inflow of people will create further pressures, thereby affecting the capacity of the state to
meet its obligation towards its own people. To address this glaring problem,
the foremost need is to recognize them legally at national and international
levels.

4. Summary

Environmental insecurity emanating from climate change and environmental
degradation have affected various dimensions of human well-being as it creates several impediments in fulfilling the physiological need for food, water,
habitat and the like. These impacts are often of non-linear nature affecting
different regions differently given their different geographical locations as
well as diverse coping capabilities. Environmental security basically exists
only when the existence of environmental goods and services that is needed
for the survival of the humanity is not threatened by any human-induced or
natural environmental assault and people have the ability to confront the
various manifestations of these threats. Figure 1.4 illustrates(Figure-1.4) the
nexus between environmental distress and security.

According to the diagram, environmental degradation and climate change
affect the human well-being at various levels. Climate-induced resource
scarcity which leads to armed clashes in some places is only one of the
threats. The economic cost due to environmental catastrophes and environmental migrations are also creating havoc on societies. Overall, the various
dimensions of human insecurity that are not militarily solved are the results
of climate distress.

In different sections of the present discussion, efforts have been made to
sew together evidence which may project how environmental change affects
different regions all over the world. There are instances that in various

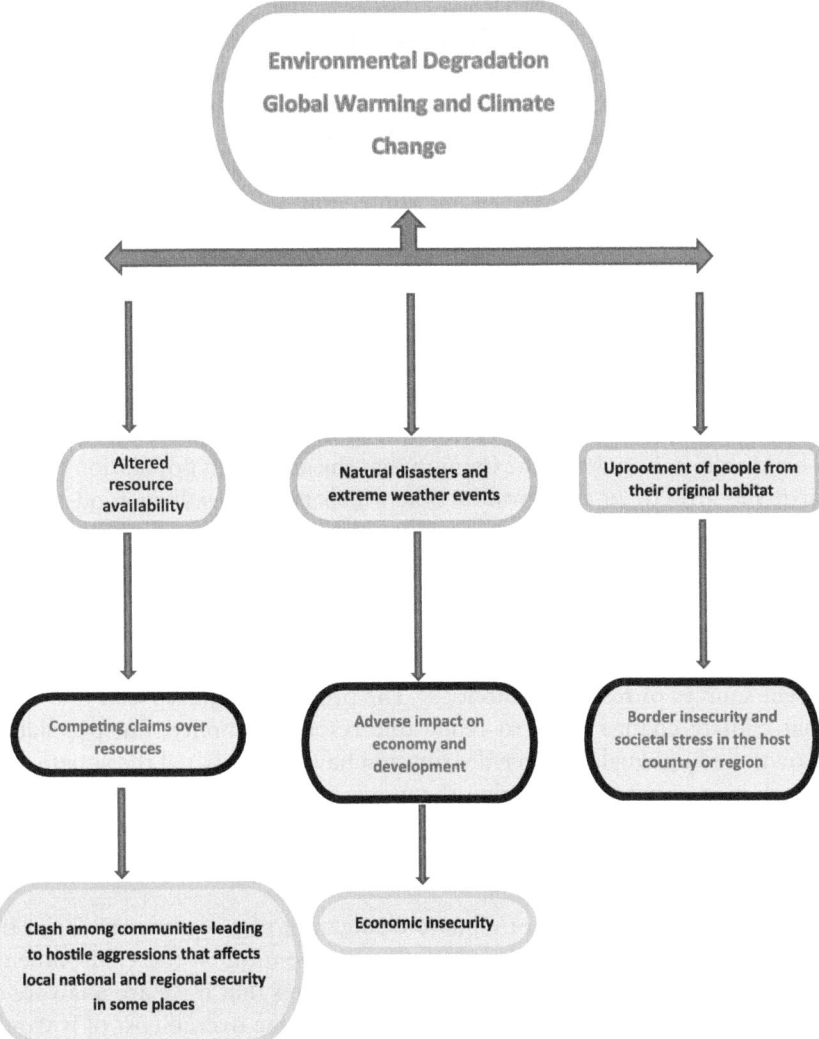

Figure 1.4 Environmental Degradation at Various Levels

Source: Prepared by the author.

volatile regions of the world resource scarcity has spurred conflicts among communities. Though they are not the soul drivers of these clashes, they can add complexity to the existing turmoil in some regions. If we try to find out the environmental routes to conflict, it becomes amply clear that adverse climatic conditions, induced by human economic activities and by their lust for unending technological advancement, ultimately alter the resource availability leading to competitive claims over these scarce resources. In the New York Times, Homer Dixon compared the severity of climate stress with that of the cold war rivalry as well as the nuclear power proliferation during the period. To him all of these affect the international security in the same magnitude.[107] Scholars underscored that the human security as a whole may be undermined by limiting the scope of access of many to life-sustaining natural resources. There may exist different reasons behind scarcity of resources; however, environmental degradation helps in speeding up the process that leads to declining agricultural yields, unavailability of potable waters, paucity of cleaner energy, deforestation and turning the croplands infertile. The magnitude of the problem is furthered when environmental scarcity induced clashes escalate the already existing conflicts whose seeds are sowed by ethno-religious tension or socio-economic inequities. In various literature highlighting resource scarcity and violent conflict, different views are presented regarding the causes of such scarcity. The Malthusians were of the opinion that the growth in population is larger than the growth of food production which might result in food scarcity owing to the demand and supply issue. Another view had been pioneered by the neo-Malthusians who believe that spike in population growth might affect the base of resources resulting in environmental destructions. As the hunger augments, there would be sufficient reasons for violent conflicts. Homer Dixon was more specific in this context while he talked about different sources of resource scarcity.[108] The project recognized that it is a fact that climate change leads to renewable resource scarcity. But population growth and unequal resource distribution have exacerbated the situation.[109] Developing countries are facing the worst due to this leading to more frequent conflict scenarios. There is thus no doubt that other reasons apart from environmental change are equally responsible for resource scarcity. In that sense, it is a risk amplifier.

The economic impacts of climate change are also significant. Since the 1990s, there is a growing trend of paying attention to the environment–economy link. The Stern report of 2006 estimates that if we are unaware of the actions required to counter climate change, the overall cost of it would be tantamount to loss of 5 per cent of global GDP every year. Considering a wider range, the estimates of damage could rise to 20 per cent of GDP or more. On the other hand, if we take necessary actions to avoid adverse impacts, it can be limited to around 1 per cent of global GDP each year.[110] Thus, environmental change and resource scarcities can take us back from where we started. But it is also true that unsustainable developmental pattern and rich Northern countries' over emissions due to energy production are responsible for the present climate crisis. But they are less vulnerable to environmental change because of their mitigation power. Their adaptive capacity is also high because of the economic affluence. But in developing countries due to population growth leading to reduce per capita availability of resources, income disparity and poverty have made the people more exposed to environmental stresses.

Given this context it is clear that today environmental degradation and climate change play such a significant role in the world politics that claims are raised to incorporate it into the modern security discourse. Environmental change as a non-conventional security threat may take various forms which increase the risk of civil violence including insurgency and ethnic conflict. Although environmental issue alone is not capable enough of triggering such violence and generally it is considered as matter of concern for low politics, sometimes the problems created by it require military preparedness to be solved. As the socio-politico and economic well-being of a country is challenged by environmental change, the nation may lose ultimately her managerial capacity as well as bargaining and coercive power. So it comes as no surprise that environment-related issues now enter the realm of high politics and international climate diplomacy has always been a tumultuous one as countries' interests never converge and each is acting to the detriment of other limiting possibilities of cooperation. For such lack of cooperative efforts, the climate apocalypse is looming large.

Notes

1 World Commission on Environment and Development, *Our Common Future*, Brundtland Report, Oxford University Press: New York and London, 1987, p.290.
2 Robert D. Kaplan, "The Coming Anarchy", in Gearóid Ó. Tuathail, Simon Dalby and Paul Routledge (eds.), *Geopolitics Reader*, Routledge: London, 1998, p.195.
3 Thomas F. Homer-Dixon, "Environmental Scarcity and Mass Violence", in Gearóid Ó. Tuathail, Simon Dalby and Paul Routledge (eds.), *Geopolitics Reader*, Routledge: London, 1998, p.205.
4 *Definition of Food Security Given by FAO at World Food Summit 1996*. Available at http://www.who.int/trade/glossary/story028/en/ Accessed on December 20, 2014.
5 Schmidhuber and Matuschke, "Shift and Swing Factors and the Special Role of Weather and Climate", in Baris Karapinar et al.(eds.), *Food Crises and the WTO*, Cambridge University Press: Cambridge, 2010, pp.137–138.
6 Bruce A. McCarl, Mario A. Fernandez, Jason P.H. Jones and Marta Wlodarz, "Climate Change and Food Security", *Current History*, Vol.112, No.750, January 2013, p.35.
7 Rupa Mukherjee, *Climate Change and Hunger*, 2019. Available at globalhungerindex.org Accessed on December 12, 2021.
8 J. Scheffran and A. Battaglini, "Climate and Conflicts: The Security Risks of Global Warming", *Regional Environmental Change*, Vol.11, No.1, March 2011, p.S31.
9 Narottam Gaan, *Climate Change and International Politics*, Kalpaz Publications: Delhi, 2008, p.308.
10 J. Scheffran and A. Battaglini, No.8, p.S34.
11 Hama Arba Diallo, "Seizing the Chance"', *Our Planet* (The Magazine of the UNEP), Vol.17, No.1, p.10. Available at http://www.ourplanet.com/imgversn/171/Hama%20Arba%20Diallo.pdf Accessed on November 15, 2014.

12 *United Nations Convention to Combat Desertification,* 1994. Available at http://www.unccd.int/Lists/Site DocumentLibrary/convention/leaflet_eng.pdf Accessed on November 15, 2014.

13 UNCCD Brochure, *Desertification, Land Degradation and Drought: Some Global Facts and Figures.* Available at http://www.unccd.int/Lists/SiteDocumentLibrary/WDCD/DLDD%20Facts.pdf Accessed on November 15, 2014.

14 *Desertification, Drought and Climate Change,* Africa Report, 2008, p.39. Available at http://www.un.org/esa/sustdev/publications/trends_africa2008/desertification.pdf Accessed on October 20, 2014.

15 A. Kannan, *Global Environmental Governance and Desertification: A Study of Gulf Cooperation Council Countries,* Concept Publishing Company: New Delhi, 2012, pp.11–13.

16 Ibid., pp.20–25.

17 Ibid., pp.34–35.

18 Julian Brger, "Darfur Conflict Heralds Era of Wars Triggered by Climate Change, UN Report Warns", *The Guardian,* June 23, 2007. Available at http://www.theguardian.com/environment/2007/jun/23/sudan.climatechange Accessed on September 20, 2014.

19 Fred Pearce, "Climate Change Is Not an Excuse for Genocide", *The Telegraph,* January 8, 2008.

20 Arab spring is a revolutionary wave of pro-democracy protests, demonstrations, civil wars in the Arab world and North Africa that started in 18 December 2010. It aimed at topple down some of the region's entrenched authoritarian regimes. Arab Spring in individual country of the region includes Jasmine Revolution (Tunisia), Egypt Revolution of 2011, Yemen uprising of 2011–12, Libya Revolt of 2011, Syria uprising of 2011–2012, etc.

21 Sarah Johnstone and Jeffrey Mazo, "Global Warming and the Arab Spring", in Caitlin E. Werrell and Francesco Femia (eds.), *The Arab Spring and Climate Change: A Climate and Security Correlations Series,* Center for American Progress, STIMSON, The Center for Climate and Security, February 2013, pp.15–18. Available at https://climateandsecurity.files.wordpress.com/2012/04/climatechangearabspring-ccs-cap-stimson.pdf Accessed on December 14, 2014.

22 "Two Dead and 400 Injured in Algeria Riots", *Al Arabiya News,* January 8, 2011. Available at http://www.alara biya.net/articles/2011/01/08/132669.html Accessed on December 14, 2014.

23 German Advisory Council on Global Change, *Climate Change as a Security Risk,* Earthscan: London, 2009, p.96.

24 J. Scheffran and A. Battaglini, No.8, p.S34.

25 German Advisory Council on Global Change, No.23, p.87.

26 FAO, *Climate Change, Water and Food Security,* Technical Background Document from the Expert Consultation, February 26–28, 2008, Rome, p.4. Available at http://www.fao.org/3/a-i2096e.pdf Accessed on February 9, 2015.

27 N. Gopal Raj, "Melting Glaciers, More Rain To Swell Himalayan Rivers", *The Hindu,* June 2, 2014, p.9.

28 Franziska Piontek, P. Michael Link and Jurgen Scheffran, *Impacts of Climate Change on the Nile River Conflict: The Case of Egypt 2020,* pp.36–38. Available at http://clisec.zmaw.de/fileadmin/user.../piontek-et-al-2010_amman.pdfa Accessed on June 25, 2012.

29 Ibid., p.39.

30 Norman Myers, "Environment and Security", *Foreign Policy,* No.74, Spring 1989, pp.28–29.

31 Shlomi Dinar, "Environmental Security", in Gabriela Kutting (ed.), *Global Environmental Politics: Concepts, Theories and Case Studies,* Routledge: London, 2011, p.65.

32 Norman Myers, No.30, p.29.
33 Kath Weston, "Water and War", in Karrie lynn Pennington and Thomas V. Cech (eds.), *Introduction to Water Resources and Environmental Issues*, Cambridge University Press: New York, 2010, pp.442–444.
34 *Forests: Climate Change, Biodiversity and Land Degradation*, Brochure Published by the Joint Liaison Group of the Rio Convention, 2008, p.4. Available at http://www.cbd.int/doc/publications/for-cc-2008-en.pdf Accessed on December 21, 2014.
35 Second Largest Rate of Amazon Deforestation in Brazilian History, *Greenpeace Press Release*, May 19, 2005. Available at http://www.greenpeace.org/international/en/press/releases/second-largest-rate-of-amazon/ Accessed on December 21, 2014.
36 Patti L. Petesch, *North-South Environmental Strategies, Costs and Bargains*, Overseas Development Council: Washington, DC, 1992, pp.32–37.
37 Donald E. Weatherbel, *International Relations in Southeast Asia: The Struggle for Autonomy*, Institute of South East Asian Studies, Rowman and Littlefield Publisher: Lanham, MD and Boulder, CO, 2009, p.277.
38 Ibid., p.281.
39 Jeanne Hyde Hecker, *Promoting Environmental Security and Poverty Alleviation in the Peat Swamps of Central Kalimantan, Indonesia*, Institute for Environmental Security: The Hague, 2005, p.5.
40 Donald E. Weatherbel, No.37, pp.279–280.
41 UNEP, *The Importance of Mangroves to People: A Time to Action*, 2014, p.6. Available at http://www.indiaenv ironmentportal.org.in/files/file/The%20importance%20of%20mangroves%20to%20people_%20a%20call%20to%20action-2014.pdf Accessed on December 16, 2014.
42 Ibid., pp.76–79.
43 Gal Luft, Anne Korin, and Eshita Gupta, "Energy Security and Climate Change: A Tenuous Link", in Benjamin K. Sovacool (ed.), *The Routledge Handbook of Energy Security*, Routledge: London, 2011, pp.44–45.
44 Christof Ruhl, "Global Energy after the Crisis", *Foreign Affairs*, Vol.89, No.2, March/April 2010, pp.70–74.
45 Bernhard May, "Energy Security and Climate Change", *South Asian Survey*, Vol.17, No.1, 2010, p.22.
46 Francois Houltart (translated by Victoria Dantree), *Agrofuels: Big Profits, Ruined Lives and Ecological Destruction*, Pluto Press: London, 2010.
47 Vrushal T. Ghoble, "The Economics of Natural Gas: Its Geopolitical Implications", *World Affairs*, Vol.17, No.2, Summer 2013, p.110.
48 Ibid., pp.115–120.
49 Antony Froggatt and Michael A. Levi, "Climate and Energy Security Policies and Measures: Synergies and Conflicts", *International Affairs*, Vol.85, No.6, 2009, p.1132.
50 Francois Houltart, No.46, p.22.
51 Antony Froggattand and Michael A. Levi, No.49, p.1132.
52 Gal Luft, Anne Korin, and Eshita Gupta, No.43, 2011, p.51.
53 *Statistical Review of World Energy*, 2021. Available at Full report – Statistical Review of World Energy 2021 (bp.com) Accessed on January 22, 2022.
54 S. Solomon, D. Qin, M. Manning, Z. Chen, M. Marquis, K.B. Averyt, M. Tignor and H.L. Miller (eds.), *Climate Change: The Physical Science Basis: Contribution of Working Group I to the Fourth Assessment Report of the Intergovernmental Panel on Climate Change*, Cambridge University Press: Cambridge, 2007, pp.7–10.
55 Norman Myers, "Environmental Refugees in a Globally Warmed World", *Bio Science*, Vol.43, No.11, December 1993, p.755.

56 Ibid., pp.754–755.
57 *Impact on Vietnam Agriculture.* Available at www.tiempocyberclimate.org/ portal/archieve/vietnam/impact5.html Accessed on September 5, 2012.
58 John Barnett, *Security in Asia: Issues and Implications for Australia,* Melbourne-Asia Policy Papers/No.9, The University of Melbourne: Melbourne, p.4. Available at http://www.greencrossaustralia.org/media/81235/asialink %20-%20climate%20change%20and%20security%20in%20asia%20-%20 issues%20and%20implications%20for%20australia.pdf Accessed on December 21, 2014.
59 Christian Bouchard and William Crumplin, "Climate Change, Sea Level Rise and Sustainable Development in Small Island States and Territories", *Journal of Indian Ocean Studies,* Vol.19, No.1, April 2011, p.37.
60 Clem Tisdell, *Global Warming and the Future of Pacific Island Countries,* Working Paper No.147, The University of Queens Land, 2007, pp.11–12. Available at http://www.uq.edu.au/rsmg/docs/ClemWPapers/EEE/WP147.pdf Accessed on December 21, 2014.
61 H. Khasnobis, "Environmental Refugees: Climate Change, Conflict and Forced Migration", *The Statesman,* April 6, 2012.
62 Clem Tisdell, No.60, p.12.
63 H. Khasnobis, No.61.
64 Justin Hoffman, *The Maldives and Rising Sea Levels,* ICE Case Studies, No.206, May 2007. Available at http://www1.american.edu/ted/ice/maldives. htm Accessed on December 21, 2014.
65 Ubydul Haque, Masahiro Hashizume, Korine N. Kolivras, Hans J. Overgaard, Bivash Das and Taro Yamamoto, "Reduced Death Rates from Cyclones in Bangladesh: What More Needs to Be Done?", *Bulletin of the World Health Organization,* October 24, 2011. Available at http://www.who.int/bulletin/ volumes/90/2/11-088302/en/ Accessed on December 15, 2014.
66 "Cyclone Sidr in Bangladesh", *The Guardian,* November 19, 2007. Available at http://www.theguardian. com/flash/page/0,,2212365,00.html Accessed on December 15, 2014.
67 Ilan Kelman, *Disaster Diplomacy: How Disaster Affect Peace and Conflict,* Routledge: London, 2012, pp.49–53.
68 Marina Kamenev, *Australia's Worst Cyclones: Timeline,* February 2, 2011. Available at http://www.australiangeographic.com.au/topics/science-environment/ 2011/02/australias-worst-cyclones-timeline/ Accessed on December 15, 2014.
69 S. Kreft, D. Eckstein, L. Junghans, C. Kerestan and U. Hagen, Briefing Paper, *Global Climate Risk Index 2015: Who Suffers Most From Extreme Weather Events? Weather-related Loss Events in 2013 and 1994 to 2013,* Germanwatch, November, 2014, pp.7–8. Available at http://www.indiaenvironmentportal. org.in/files/file/global%20climate%20risk%20index%202015.pdf Accessed on December 16, 2014.
70 Eckstein David, Künzel Vera and Schäfe Laura, *Global Climate Risk Index, 2021: Who Suffers Most from Extreme Weather Events? Weather-Related Loss Events in 2019 and 2000–2019 Briefing Paper,* 2021. Available at https:// germanwatch.org/sites/default/files/Global%20Climate%20Risk%20Index%20 2021_1.pdf Accessed on January 22, 2022.
71 Ilan Kelman, No.67, pp.40–41.
72 Article 1 of the 1951 Geneva Convention defines a refugee. *Convention relating to the status of refugees,* Art.1, July 28, 1951. Available at http://www.unhcr. org/3b66c2aa10.pdf Accessed on December 25, 2011.
73 As quoted by Astri Suhrke, *Pressure Points: Environmental degradation, Migration and Conflict,* Paper prepared for the workshop on "Environmental Change, Population Displacement, and Acute Conflict," held at the Institute for

Research on Public Policy in Ottawa in June 1991 as part of the "Environmental Change and Acute Conflict" project of the Peace and Conflict Studies Program, University of Toronto and the American Academy of Arts and Sciences: Cambridge, MA, p.6. Available at http://www.cmi.no/publications/file/1374-pressure-points-environmental-degradation.pdf Accessed on December 21, 2014.

74 Oli Brown, IOM Migration Research Series, *Migration and Climate Change*, No.31(2007)IOM, Geneva, 2008, p.15. Also available at http://www.iom.int

75 Quoted by Angella William, "Turning the Tide: Recognising Climate Change Refugees in International Law", *Law and Policy*, Vol.30, No.4, October 2008, Baldy Center for Law and Social Policy, p.511.

76 Oli Brown, No.74, p.18.

77 Laura Story Johnson, "Environment, Security and Environmental Refugees", *Journal of Animal and Environmental Law*, Vol.1, 2009–10, p.235. Available at https://drive.google.com/file/d/0B0gcImiUSq5Ebm9TZ1VwZlZzd2s/view?pli=1 Accessed on November 16, 2012.

78 Norman Myers, "Environmental Refugees: A Growing Phenomenon of the 21st Century", *Philosophical Transactions of the Royal Society of London Series, Biological Sciences*, Vol.357, No.1420, p.609. Available at http://www.ncbi.nlm.nih.gov/pmc/articles/PMC1692964/pdf/12028796.pdf Accessed on December 21, 2014.

79 Quoted in, J. Lovel, "Climate Change to Make One Billion Refugees Agency", *Reuters*, May 13, 2007. Available at http://www.reuters.com/article/latestcrisis/idusl10710325 Accessed on December 21, 2014.

80 Bassetti Francesco, *Environmental Migrants: Up to 1 Billion by 2050*, May 22 2019. Available at Environmental Migrants: Up to 1 Billion by 2050 – Foresight (climateforesight.eu) Accessed on January 22, 2022.

81 Norman Myers, *Environmental Refugees: An Emergent Security Issue*, Paper Presented at the 13th Economic Forum, Prague, May 2005, pp.1–2. Available at http://www.osce.org/eea/14851?download=true Accessed on December 15, 2012.

82 IDMC, *Global Report on Internal Displacement*, 2021. Available at internal-displacement.org Accessed on December 12, 2021.

83 Astri Suhrke, No.73, pp.16–19.

84 International Organisation for Migration, *Migration, Climate Change and the Environment*, Compendium of IOM's Activities, p.16. Available at http://www.iom.int/jahia/webdav/shared/shared/mainsite/activities/env_degradation/compendium_climate_change.pdf Accessed on December 21, 2014.

85 IOM, *Internal Displacement in the Context of the Slow Onset Adverse Effects of Climate Change*, 2020, pp.8–10. Available at https://www.ohchr.org/Documents/Issues/IDPs/International-Regional/iom-idp-climate.pdf Accessed on January 22, 2022.

86 IOM Policy Brief, *Migration, Climate Change and the Environment*, May 2009, p.3. Available at http://www.iom.int Accessed on December 22, 2014.

87 Ibid., p.4.

88 Oli Brown, No.74, p.23.

89 News and Press Release, *Somalia Braces for Record Levels of Displacement as Drought Takes Hold*, April 21, 2021 Available at https://reliefweb.int/report/somalia/somalia-braces-record-levels-displacement-drought-takes-hold Accessed on December, 15, 2021.

90 IOM, No.85.

91 A.K.M. Abdus Sabus, "Disaster Management System in Bangladesh: An Overview", *India Quarterly*, Vol.68, No.1, March 2012, pp.32–35.

92 Oli Brown, No.74, p.24.

93 International Organisation for Migration, No.84, pp.73–148.

94 A.K.M. Abdus Sabus, No.91, pp.31–32.
95 International Organisation for Migration, No.84, pp.221–222.
96 IDMC, *Global Report on Internal Displacement*, 2021. Available at https://www.internal-displacement.org/global-report/grid2021/ Accessed on December 23, 2021.
97 IOM Policy Brief, No.86, p.4.
98 Norman Myers, No.78, p.611.
99 Angella Williams, No.75, p.505.
100 Quoted in DB Jagmohon. *Thengadi Memorial Lecture 2007 on Crisis of Environment and Climate Change*, Allied Publisher: Mumbai, 2008, p.4.
101 International Organisation for Migration, No.84, p.81.
102 Angella Williams, No.75, p.515.
103 IOM Policy Brief, No.86, p.4.
104 Astri Suhrke, No.73, p.7.
105 Ibid., pp.8–9.
106 Adrian Martin, "Environmental Conflict between Refugee and Host Communities", *Journal of Peace Research*, Vol.42, No.3, May 2005, pp.335–337.
107 Quoted in Idean Salehyan, "From Climate Change to Conflict? No Consensus Yet", *Journal of Peace Research*, Vol.45, No.3, May 2008, pp.315–316.
108 Henrik Urdal, "People Vs. Malthus: Population Pressure, Environmental Degradation and Armed Conflict", *Journal of Peace Research*, Vol.42, No.4, Special issue on the Demography of Conflict and Violence, July 2005, pp.418–420.
109 Thomas Homer-Dixon, "Environmental Scarcities, State Capacity and Civil Violence", *Bulletin of the American Academy of Arts and Sciences*, Vol.48, No.7, April 1995, p.26.
110 Nicholas Stern, *The Economics of Climate Change: The Stern Review*, January 2006, p.iv. Available at http://www.hmtreasury.gov.uk/independent_reviews/stern_review_economics_climatechange/stern_review_report.cfm Accessed on December 22, 2014.

2 Effects of Environmental Change and Its Negative Consequences in India

Environmental change affects different regions in different ways. Although it is true that it causes scarcity of natural renewable resources, leads to sea level rise, creates severe storms, droughts and floods as well as alters agricultural production across the world, this threat can more relevantly be defined in terms of national vulnerabilities. Various reports have underscored the security implications of environmental problems. Same was echoed by the Climate Security Report released by the American Security project on 1 November 2012. The report emphasized the role of climate change contributing towards local and regional conflicts in the hot zones of the world. Climate change has exacerbated the problems of overpopulation and resource scarcity.[1] Since the environmental issue has gained such universal currency, its acute state in India cannot be ignored. Here, also the green dimension of politics has emerged as a significant policy concern at all levels.

1. Resource Scarcity in India

Food, water, forestry and energy resources are vital to any country's economy, and the same scenario is present in India. The natural resource base of the country is strained due to population growth most of whom are dependent on climate-sensitive sectors like agriculture and forestry. It is estimated that the environmental cost in India is more than 10 per cent of the gross domestic product. Loss of farming output, water scarcity and deforestation result in such scenario.[2] Such climate-induced resource scarcity combined with other hostile situations like rising sea levels, severe weather events as well as exodus of climate refugees affect not only the Indian economy but also its people, its political relationships with neighbours, infrastructure and ultimately her national security and interests.

1.1. Food Scarcity and Land Degradation

The agro-based economy of India is severely impacted by rapid environmental change, resulting in acute food scarcity. A study by William R. Cline shows that India and Pakistan will be among the worst affected of the 29 developing countries losing nearly 20 per cent of farm output as a consequence of global

DOI: 10.4324/9781003271192-3

warming. India's agricultural output could fall under standard assumptions by 31–37 per cent in the southeast and southwest, and by 58–61 per cent in the northeast and northwest. Pakistan could also lose 20 per cent or more of its food production.[3] The IPCC 4th Assessment Report also warns that the most populous regions of the world will face the brunt of climate disaster more, and South Asia in particular will largely suffer as new threats emerged to its strategic resources like water, food and energy.

India, which has 35 agro-climatic zones, is highly exposed to climate shifts. Higher temperature, changed rainfall patterns and sea level rise induced decline in farm output and the consequent food insecurity are the results of changing climate here. The rain-fed agriculture is particularly affected by the adverse impact of weather, especially by the timing and duration of the rainy season, the total amount of rain received and the pattern and severity of storms.[4] The reason is that the country has states like Punjab and Haryana with assured irrigation, while more than 40 per cent of the harvesting area is only rain dependent. Global climate change therefore affects the agricultural output vastly. A grim situation detrimental to a good farm output appeared in the year 2012, when the country as a whole had received rainfall 21 per cent below average.[5] Table 2.1 projects how monsoon-dependent Indian agriculture has suffered in the year 2012.

Table 2.1 The Grim Situation of Agriculture in 2012 Across India

Region	Rain-Dependent Area in percentage	Situation in 2012
Southern Peninsula (Karnataka, Andhra Pradesh, Kerala, Tamil Nadu)	90	Late and scanty rains have affected transplanting of rice across 19.72 lakh hectares of land.
Central India (Chhattisgarh, Gujarat, Madhya Pradesh, Maharashtra, Orissa)	70	Scanty rains have affected sowing of rice, sugar, cotton, coarse cereals affecting cropping 4.43 lakh hectares of land.
East and Northeast India (Assam, Andhra Pradesh, West Bengal, Bihar, Jharkhand)	95	Heavy rains have flooded the plains across 23 districts in Assam, Bengal and Bihar affecting 3.3 lakh hectares of farm cropping.
Northwest India (Haryana, Himachal Pradesh, Punjab, Uttar Pradesh, Uttarakhand)	20	Late and scanty rains have affected the growth of staples such as rice, sugar, cotton and coarse cereals. Deficient rains have also meant that water reservoirs vital for irrigation and power have hurt rice transplantation in 6.55 lakh hectares of land in Punjab, Haryana and Uttar Pradesh.

Source: Data collated from Zia Haq and Gautam Choudhury, "India's Mid-Summer Nightmare", *The Hindustan Times*, July 29, 2012.

However, not only water scarcity but rising temperature also contributes to low agricultural output in India.[6] Not only that India is suffering from land degradation and desertification hugely, but Indian agriculture also suffers from desertification and degradation. Around 25 per cent of Indian land is affected by desertification, while 32 per cent of it is suffering from degradation. The estimated total of 328 million hectares of geographical area in India of drylands constitutes an area of 228.3 million hectares (about 69.6 per cent). Desertification, land degradation and drought (DLDD) have hit 7,91,475 sq. km of the territory covering almost all states and union territories of the country.[7] According to India's Fifth National Report on Desertification, the most desertified land is found in Rajasthan which was followed by states like Gujarat, Maharashtra in the West, Jammu and Kashmir in the North, Orissa in the East and Andhra Pradesh in the south.[8] As a result, food production is seriously threatened that results in food insecurity for millions of Indians. Deforestation, unsustainable extraction and use of forest products, overgrazing, shifting cultivation, unscientific crop rotations, inappropriate irrigation, depletion of groundwater, urbanization and climate change induced loss of vegetation covers are the most obvious causes of land degradation and desertification in India.

While agriculture is severely affected by the adverse climate cataclysm, the former is also responsible for polluting the environment. Indian estimates show agriculture accounts for 17 per cent of GHG emissions in 2007 and even more if emissions associated with inputs like fertilizers and distribution are taken into consideration. CH4 which is a GHG is emitted from agricultural sector due to enteric fermentations and rice paddy cultivation. Another GHG N_2O is also emitted because of the use of fertilizers.[9] Along with this agriculture-induced emission, the rising demands for more food due to demographic expansion can take a heavy toll on the environment. Increased food production may degrade other natural resources in many ways. The natural resources needed for the agricultural expansion are also subject to degradation. Conversion of forestlands for agriculture and biodiversity loss and soil erosion may also result from this, thereby pushing agriculture against the environment.

Diversification of land for nonagricultural purposes and promotion of export-oriented cash crop production with monoculture in many parts of India have further trimmed the food output. Large-scale displacement follows from this with ill-effects on the economy and society. Adopting the industrial agriculture and livestock breeding in India has also had serious consequences for the survival of biodiversity. This is evident in the fact that in 1950, there existed 30,000 which was declined to just over a thousand varieties in 2012. Such evidence has indicated the loopholes in policies that are acting to the detriment of our basic capital.[10] Food security which largely depends on the diversity of crops is largely threatened by the commercialization of seeds and by the introduction of FDI in retail market too. The corporate giants basically try to gain profit utilizing the genetic wealth of India at the cost of local farmers' interest.

However, in India, where almost 70 per cent of the workforce has engaged themselves in agricultural activities for employment and subsistence, the governmental response to this severe agrarian distress is often lopsided. The problem has further been accelerated as the strategies required for achieving food security have its own limitations. For instance, in our countries, often the policymakers have favoured the use of biotechnology in agriculture for achieving food security. But various stakeholders involved differ in their opinion about the positive and negative impacts of such transgenic produce on agriculture. The genetically modified version of rice, brinjal and mustard might come in fierce competition with the locally grown and wild varieties of these crops.[11] One of the assumed benefits associated with GM crops that they are of high yielding nature does not have any proper evidence. There are also lack of sufficient bio-safety test before their release as was the case of the commercial release of BT brinjal. Moreover, the entering of GM components into the food chain causes detriments to the environment. Along with such dire consequences, the capital-intensive nature of such crops leads to poverty and intensified debt pushing millions of farmers to commit suicide.[12] So various measures to augment food production have their own shortcomings which actually contribute to further human misery. At this juncture, what we need is an appropriate adaptation and mitigation strategy, well-founded research and development activities to promote sustainable farming with drought-resistant, salt-tolerant crop varieties as well as efficient use of bio-technology for organic farming. Without such efforts, food insecurity could increase and may lead to heightened risk of undernourishment and hunger.

Food security of any country depends on many factors as is evident in India. Often the ill-effects of climate change coincide with political and economic marginalization too. Corruption, financial poverty, faults in the public distribution system and commercialization of food are also major actors in the acceleration of the problem at large. While climate-induced poor farm output has already affected the population and economy gravely, non-recognition of the fact that most of our population lacks the basic resources, both monetary and non-monetary to get access to a sufficient quantity of food, has exacerbated the situation. Scientific understanding of the problem has not also translated into a focused policy response. So it is futile to blame the climate-triggered decreasing productive capacity per se rather than the economic and political structures and social dysfunction that are essentially human constructs and that control food allocation process in our society. The problem of scarcity of resources like food, so, can only be checked if there is an overall amendment to the existing agricultural policies. Such radical changes can obliterate problems of hunger and inequality in resource distribution and might promote sustainability.

1.2. Water Scarcity

The scarcity of food and water are closely linked as poor harvest requires more water whose availability has also been strained due to environmental change. Changes in environment impact the hydrological cycle due to strained surface water level and reduced groundwater recharge. But the severity of the problem as well as the vulnerability to it differs across regions. According to the IPCC Fourth Assessment Report, due to climate catastrophe, the pattern of precipitation would change disproportionately leaving subtropics more drought-prone and susceptible to disruption in rain-fed farming, and to dwindling supply of water required for industrial, household and agricultural use.

The Ministry of Environment and Forests (MoEF) of Government of India published the Initial National Communication to the UNFCCC which describes how water resources are affected by climate change. The report describes how sea level rise and glacier melting adversely affects the water cycle of India. Coastal inundation and associated salinization affects the groundwater quality of the coastal zone of the country as well.[13] Both extreme climate-induced drought and floods pose a threat to water resources, and thus perennial scarcity of potable and usable water in India threatens the human lives and livelihoods. It is estimated that in India, climate change leads to remarkable reduce in per capita water availability by 2025.[14] A different study noticed that water stress would increase even before 2025 when per capita availability of water would be fallen below 1,000 m³. Climatic and demographic factors are responsible for that.[15]

In India, the groundwater quality in the alluvial aquifers is affected by the increased frequency and severity of floods. This may lead to severe drought conditions affecting watercourses. Sea level rise induced saltwater intrusion into coastal fresh water systems is also a matter of great concern. Rising salinity due to sea water intrusion affects hugely the groundwater level in Tamil Nadu and Gujarat. Arsenic has been contaminated with the groundwater in West Bengal and Orissa endangering human lives. This happens due to the overuse of groundwater that causes fall in water table.[16]

Climate change induced drought results in higher demand for water in many parts of India. As a consequence, unsustainable extraction and overuse of groundwater including the reserve of non-renewable ones follow that actually escalate water stress in many regions of the country.[17] This is to note that recharge to groundwater and flows to surface storages are impacted by climatic changes as a result of changing flow patterns, increasing temperatures, increasing losses, increasing extractions, water available for recharge and that available in the catchment of surface storages.[18] It has been projected that by 2040, water stress in India would be escalated at an alarming rate.[19] Increasing urbanization and industrialization as well as population rise have further worsened the situation as they directly and indirectly affect the quality and demand and supply of this life-sustaining resource. For instance, the non-renewable groundwater in the state of Andhra Pradesh, Gujarat, Haryana, Punjab, Tamil Nadu and Maharashtra is depleted fast as a result of increasing urbanization and unregulated industrialization. Moreover, nonagricultural extraction renders the traditional base of surface water sources weak, encouraging indiscriminate groundwater extraction, while new ones are not being created. One well-studied example of this trend is Bangalore, which now has only 14 live lakes out of 51 in the 1970s. Water bodies outside many cities are pathetically being turned into sewage ponds because sewerage infrastructure inversely related to the pace of urbanization.[20]

Primarily climate change is responsible for water stress but competing demand for water from different sectors like industry, farming as well as different water needs of urban and rural users escalates the paucity as well. Nearly 15 per cent of food production requires groundwater leading to depletion of aquifers. The densely populated areas are suffering more from this.[21] The lack of sustainability in managing this resource has further augmented the problem.

As agriculture forms the backbone of the rural economy of India, sometimes protests came from farmers on the ground that water meant for agriculture is being diverted to industry or urban areas indiscriminately, thereby preventing them to make access to their entitled water. Such was the root of the protests against the Sophia power project in Amravati, which requires 87 million cubic metres from the upper Wardha dam. It was alleged that if this diversion takes place, nearly 23,219 hectares of land will be deprived of irrigation.[22] Sometimes water scarcity is the byproduct of irresponsible government policy and unsustainable agriculture too. In India after green revolution, water productivity has been ignored and the focus is shifted to producing water-intensive crops, which ultimately leads to reduction in the moisture conservation capacity of the soil. Before the advent of green revolution, water conservation was an intrinsic part of indigenous agriculture. But after green revolution irrigation system had started replacing rainfed agriculture. Use of chemical fertilizers and the spread of monoculture affected the soil leading to recurrent soil moisture drought.[23] The green revolution introduced massive irrigation projects and water-intensive farming like wheat and rice production which add more water to an ecosystem than its natural drainage system can accommodate, resulting in water logging, salinization and desertification. Such water logging at the Malaprabha irrigation project in the Krishna Basin led to the farmer rebellions. The sudden climate change, the intensive irrigation and cultivation of cash crops like cotton aggravated the problem. Farmers were shot by police when they refused to pay water taxes which was levied by irrigation authorities as betterment levy. The protests rapidly spread to other parts of Karnataka, and finally the government ordered a moratorium on the collection of water taxes.[24]

GM food crop yielding has often been suggested by the government as an alternative to water-intensive farming, but such mechanism has its own limitations and sometimes has led to soil erosion. Climate change augments all these in the long run. So, in many villages of India along with failed monsoon and resultant droughts, poor water management system, forced cultivation of water-intensive crops for commercial gains as well as class- and caste-based resource (land, water, forest) distribution have paralysed the agro-based economy. So it is evident that climate-induced water crisis is aggravated by inappropriate and profligate water policy. The end result is migration, social unrest and inter-community conflict distorting the socio-political fabric of India.

Climate change has a tremendous effect on glaciers. The acceleration of melting of glaciers is likely to cause increase in river flows that initially results in higher incidence of flooding and landslides. But later as the volume of ice available for melting decreases, the glacial runoff and river flows are expected to get diminished.[25] Bhutan, Nepal, Bangladesh and India would have experienced serious consequences due to disappearance of glaciers and frequent glacial lake burst. This will initially cause increased flooding and mudslides followed by an eventual decrease in flow in rivers that are gla-cier fed. World Wide Fund for Nature projection indicates that the annual renewable Ganges flow could slip into a situation of scarcity by 2025. This could further jeopardize supply particularly in the dry seasons prior to the monsoon.[26] The complexity and variability of river flow generation are also vulnerable to such effect.

The Himalayan mountain system which is greatly susceptible to global warming is also a matter of great concern here. It has not only colossal water resources but also the frontline of global warming. The greater Hima-layas that have been called the earth's Third Pole as they store the third larg-est volume of fresh water (the 15,000 Himalayan glaciers cover 30,000 sq. km or 17 per cent of the mountain area and hold 12,000 cubic kilometres of fresh water), comprising the Hindu Kush, the Himalayan range and the Tibetan plateau, are warming at a rate that is two or four times higher than the global average, and estimated at 2–2.4 degree centigrade.[27] As a result and because of the specific effect of black carbon, Himalayan glaciers are melting at unprecedented rates, the snow season is shortening and the snow line is moving to higher elevation. It would result in dwindling water flows in some of the major rivers like Ganga, Indus, Brahmaputra and Yangtze having negative impacts on India. Such climate change induced dwindling glacier water flows not only affect the irrigation and household activities of people of this region but also undermine the energy potential of all the hydroelectric power projects on the Himalayan Rivers.

Along with glacial shrinkage and retreat, climate change has largely impacted distribution of precipitation over time and space leading to both drought and flashfloods.[28] The situation in India is more critical due to its geographical location. Being a middle riparian, we are in demand of water from the upstream countries while having the obligation to maintain the proper flow required by the downstream countries.[29] Thus, as a river runs through more than one country, the countries concerned must reach an agreement on the sharing of the waters of that river. The main rivers of the Indian subcontinent are the Indus, the Ganga and the Brahmaputra. Both the Indus and the Brahmaputra have originated in Tibetan Himalayas. The

Tibetan Plateau is also experiencing faster glacial melt and other ecological change as well as global warming induced shrinkage of its permafrost that ultimately depletes the water resources – 'a lifeline for the peoples of several densely populated Asian Countries'.[30] The river Ganga is also threatened by climate change as the Gangotri Glacier from which river Ganga is originating is experiencing shrinkage.[31] So apart from the close interconnection between food and water scarcity induced problems, another issue of great concern is shared water resources whose flows are also affected by the climate change. As a result, it has escalated the intensity of already strained riparian relations between India and her neighbours as they differ in their claims to a 'reasonable share of water'.

In relation to shared water resources in the Ganga–Brahmaputra–Meghna basin, Ramaswamy R. Iyer has classified Nepal as upper riparian, India as middle riparian (Lower riparian to Nepal, Bhutan and Upper riparian to Bangladesh) and Bangladesh as lowest riparian deltaic country.[32] Against such a geographical scenario often efforts for required augmentation of both inter- and intra-state river water in the lean seasons through diversion of surplus of other rivers as well as dam building over rivers which leads to forced resettlement of population as well as extreme climate variability may act as drivers for conflict escalations in the subcontinent. There are three fundamental treaties – The Indus Treaty of September 1960 between India and Pakistan, The Ganga Treaty of December 1996 between India and Bangladesh, and the Mahakali Treaty of February 1996 between India and Nepal, which have tried to resolve the discord and blame game over dam building and impounding of shared river water. Although these treaties have gained international recognition as successful instruments of conflict resolution, the signatories have still witnessed crisis over the entitlement of shared water resources. The riparian relations between India and her immediate neighbours make it amply clear that the increased likelihood of water conflicts in Asia, to a large extent, is attributed to higher dependency on cross-border river inflows which are determined by changes in climate to some extent.

India has strained relations with China also in this regard. China, being an upper riparian country and having control over Tibetan plateau, in contrast to other countries of this region, is in an advantageous position as minimum percentage of its water resources comes from across its border. Further, it requires huge water for its large population as well as for industrial use. So it is in her national interest to augment the volume of useable water. China is therefore accused and suspected of pursuing major water transfer projects both inter- and intra-basin on the Tibetan plateau[33] which may affect the water flows to downstream countries including India.[34] Specifically, diversion of the Yarlung Tsangpo branch of the Brahmaputra around the famous U-bend before the river enters India is a matter of concern for the country. India also blamed China for initiating a hydropower project at Zangmu in Tibet, along the Yarlung Tsangpo having adverse effects downstream which was denied by China on the ground that it does not involve substantial diversion of a river's waters, thereby not significantly impacting areas in India downstream.[35]

The water diversion project by China affects the natural landscape of Northeastern part of India. In the East Siang District of Arunachal Pradesh, a territory administered by India but claimed by China, people in the town of Pasighat reported that the usually strong river Brahmaputra suddenly dwindled to almost nothing in early 2012. The state's Minister of Water Resources demanded an investigation into whether the shortage had been caused by dam building upstream on the Chinese-controlled portion of the river.[36] Although it is not clear that what caused the sudden but temporary cessation of the Brahmaputra's flow in March 2012, the political distrust between the two countries has made it clear that future water conflict over shared water resources like the river Brahmaputra is very much likely to take place. Environmental disaster due to the diversion of the river is another matter of concern. The Great Bend is located in an area of high seismic activity close to the fault line where the Indian plate collides against the Eurasian Plate. This might lead to severe earthquake in the region.[37]

Water is a crucial medium of conflict and cooperation. When shared water resource is the case in point, competition for access to it has exacerbated the situation. The changes in the 'physical or political setting' of international river basins by 'construction of a dam upstream, diversion for irrigation purposes, or realignment of political frontiers' are responsible for fostering competition among co-riparians. In this context, Ismail Serageldin's much quoted prediction is worth mentioning – the future wars will be guided by water only. Although there is no such instance of overt war fought simply over water, this life-sustaining resource has been an underlying factor in several armed conflicts. Climate change has also exacerbated the problem by decreasing the flows of snow-fed river waters and by altering the quantity of available water necessary for irrigation and household use as well as minimizing the hydropower generation capacity of these rivers.

Such discontent over dam building is not restricted to trans-border rivers only; within India, irreconcilable water conflict has also erupted among states that disagree over ownership and distribution of water. For instance, the Kaveri River dispute that has given birth to water war between Tamil Nadu and Karnataka and both the countries have witnessed bloodshed and fall of governments over this issue. So water is such a medium through which climate change affects the humanity severely. India, as not an exception, has witnessed scarcity of this resource due to adverse climate variability which not only reduces her per capita fresh water availability but also increases the intensity and frequency of rain and flashfloods and sea level rise that determine this availability. All of these ultimately lead to severe degradation of ecosystems including decrease in crop yield hampering the food security of the country. Along with natural calamity, the man-made disasters as well as antagonistic political relationship between India and her neighbours have accentuated the problem.

1.3. Deforestation

India is well known for its rich biodiversity with a large variety of flora and fauna. It has multiple climate zones ranging from the great Himalayas and the Ghats to the northeastern rainforests. As many as 16 varieties of major forests, 7 per cent of world's biodiversity is found here. The rich variety of forests includes tropical, subtropical forests, alpine pastures and also mangroves.[38] Many communities live in and around the vast forest land, and forest provides life-sustaining resources to them. Non-timber forest products are found in Indian forests in a very significant quantity. So deforestation has heavy economic and social implications for India.

While analysing the impact of climate change on forest, Prof. Ravindranath estimated that nearly 75 per cent of forests in India would witness changes by the end of this century. The changes include their net primary productivity and biodiversity.[39] In his study of finding the degree of deforestation in Ghats, it has been noted that Uttar Kannada district lost around 2000 hectares of forest cover, while Shimoga lost 4,000 hectares during the same period. Another study noted that the vulnerability of these districts would increase due to the projection of climate change even before the end of 2030.[40]

Climate change affects the coastal plains of India also. The mangroves are at severe risk due to storm surges, SLR-induced coastal inundation. The projected change in the climate might affect the Sundarbans with acute effects on the wild habitats reside there. A study reveals not only the coastal area but also the Himalayas would witness severe deforestation by 2100, resulting in denudation of about 90 per cent of dense forest.[41] In a 100-year projection of climate change a study calculated a loss of 32 per cent of forest cover in the Himalayas.[42] As forests are natural dams conserving water in catchments, such deforestation along with commercial forestry also make the extreme weather events more dreadful. The State of Forest Report 2021 has shown a positive scenario where it has been noted that there is an increase of over 2,000 sq. km in the country's total forest cover. However, this is not a uniform trend. The forest cover of the Northeastern region has witnessed an overall decline of more than 1,000 sq. km. Though the carbon stock has increased in 2021 as compared to 2019, the report estimated that by 2030, over 60 per cent of forests in India will be affected.[43] In such a scenario, the fundamental need is to find out the reasons behind vanishing forest lands and their degradation aggravating climate disorder in India. The reasons may be the following:

1 Diversion of forestlands for non-forest use and damage caused by mining, irrigation projects, industries, roads and shifting cultivation;
2 Conflicting claims by different sectors of the society to assert control over forest and forest-derived goods and services;
3 Lopsided forest governance system and national response to international move against deforestation.

The changing land use pattern to feed the rising population of the country has serious bearing on forests. Large area of forests has been converted into agricultural lands. To fulfil the need for fuel and places for human settlement, forests are destroyed at an alarming rate. Almost 13 million hectares of forest land across the world are cleared every year, a report suggests. In India, similarly the forestry sector is impacted by other sectors such as agriculture, energy and industry. Several hectares of forest land of the country are converted due to traditional shifting agriculture to feed the burgeoning population. For instance, the loss in forest cover in the Northeastern states was mainly due to the shifting cultivation. Woods are also destroyed for energy consumption. In India, 85 per cent of fuel use comes from biomass which was collected from our forests. It affects the latter's existence also.[44] Another instance of forest degradation is relevant here, that is, significant mangrove losses due to aquaculture which are specifically visible on the west coast of India and in the districts of Karwar and Jumta in Karnataka state, Palghar and Shrivardhan in Maharashtra, and Valsad I in Gujarat. Issukapalli mangrove forests, which once stretched 500 hectares in Andhra Pradesh, have been reduced significantly. All over India where once stood mangrove forests now lie roads and aquaculture ponds. In Orissa and West Bengal, several shrimp farms have been established in mangrove woods that are lucrative for shrimp growth. In the Sundarbans of Bengal, shrimp ponds have been constructed on 35,000 hectares of land once inhabited by mangrove forests.[45] Such export-driven shrimp farming throughout the coastal regions of India has decreased the coastal zone buffer capacity and left the regions vulnerable to cyclones and floods and new range of environmental disasters. It is claimed by the experts that the destruction caused by the super cyclone in Orissa could have been avoided or reduced had the mangroves along the coastline not been destroyed for shrimp farming.[46] Excess use of timber and pollution are also creating problems in the Sundarbans.

Not only the mangroves are destroyed by human activities, in other parts of the country, but enterprise-driven deforestation is also rampant. In India, there are instances that thousands of hectares of forests have been destroyed by the building of hydroelectric dams. Such incidents are often opposed by the local communities as forests provide them with fuel, food, medicinal herbs and fodder – while maintaining an ecological balance. It is well known how in the 1970s, the commercial deforestation of the fragile Himalayan ecosystem had vehemently been opposed by the local community which gave birth to the famous 'Chipko Movement'. The tribal areas of central India had also witnessed escalating conflicts between villagers and the forest administration in Bihar, Orissa and central Indian states like Madhya Pradesh, Maharashtra and southern states like Andhra Pradesh during that period.[47] The silent valley movement remains another instance of people's resistance to environmentally destructive projects. When the Kerala State Electricity Board announced plans to begin the establishment of a 240-megawatt hydroelectric project over the Kunthipuzha river flowing through Palakkad and Mallapuram districts in 1976, mass protest was triggered by overthrowing the project that would submerge a huge patch of the virgin evergreen forests rich in biodiversity and sealing all corners of the

Valley from developmental projects.[48] The Narmada Valley Hydro Project also faced resentment for submerging large areas of forest and cultivated lands leaving dangerous environmental distress. Not only the hydroelectric projects, for other reasons, but also forests are destroyed and degraded in India. The Renuka storage dam that is to be constructed in Sirmour district of Himachal Pradesh for solving the drinking water shortage problem in Delhi has spurred another controversy. The dam is supposed to bring development to the backward region of Giri valley. Forty per cent of the land required for the project is forestland of which 49 hectares belong to the Renuka Wild Life Sanctuary. There were several conflicts of opinion between the HP state government and MoEF as around 1.5 lakh trees would be submerged by the project. The more complex aspect of the forestland mess is the involvement of more than 450 hectares of private forests under the category of Shamlat Lands. While the forest clearance for the project hangs fire, the environment clearance also stands challenged in the National Green Tribunal by the residents of one of the project-affected villages.[49] Unscrupulous mining in the vulnerable regions of India also intensifies deforestation. The forests in the sacred Ganmardhan Mountains which are a refuge for various plants and provide water to 22 streams which in turn fill major rivers witnessed such degradation. In 1985, Bharat Aluminium Company (BALCO) began the desecration of these sacred grounds which faced severe obstruction by the tribals of the region. But the protestors often suffered punishments and brutal atrocities by the police.[50] Similarly, Asia's largest Sal forests, situated in Mahan, Madhya Pradesh are under threat of a complete wipeout. This place is lucrative because of the vast coal reserves under the forests which have attracted corporate giants like Essar and Hindalco since 2006. The project was opposed by many given its adverse effects on the livelihood of more than 50,000 people and for it would destroy large forest area with negative impact on the wildlife.[51] The decision to go ahead with this project also met with several political blame games. The process of environmental clearance for the project was also questioned. Such hydroelectric and mining projects are built at the expense of environment. The clearance process is also flawed. In most of the cases, no detailed opinion about the pros and cons of the project or any specific recommendation from the nodal officer is attached to the approval of the projects. The role of Green Tribunal is also questioned.[52]

The forests are opened for further exploitation when on 6 February 2013, a circular was published by the MoEF mentioning that it would not be mandatory to acquire prior approval of the Gram Sabhas before conversion of any forest which was under its purview. The relaxation in the list of projects for which forest lands could be transformed has exacerbated the situation. There are 13 categories of infrastructures like schools, hospitals, irrigation canals and many more for which up to 5 hectares of forest land could be converted.[53] By removing the power from the Gram Sabha, the MoEF has basically snatched the right of the real custodian of forests to decide its fate. These approved so-called linear projects affect not only the plant and animal life in forests but the biodiversity and the people living in the forests are also affected destroying both the social fabric and ecological balance of the region.

Although it is widely accepted in India that protecting forests is significant in combating climate change, the ambiguous forest governance has further made this resource more vulnerable and fragile. Even the Indian Forest Act 1927 and Forest Policy 1988 do not clearly define the term 'forest' properly, and it has been amenable to various interpretations. In 1996, the Supreme Court of India declared that any diversion of forest land, irrespective of ownership, would require Central government's prior approval. In 2007, the MoEF reconsidered the definition and included any area under government control, notified or recorded as 'forest' under any act, for conserving and managing ecological and biological resources. It only accommodated the earlier legislations and forms of state control over forest.[54] But various interpretations of the definition ultimately make it easier to divert forest land for non-forest uses. As a result, ecological balance was at stake endangering both the forest-dependent rural livelihoods and biodiversity.

Although climate change escalates deforestation in India, such problem is accelerated due to inadequate public awareness regarding its contribution to livelihood security. The Indigenous community living on forests is also blamed for their misuse of the goods and services of this resource as wood will continue to be the major source of fuel in rural areas. Overexploitation of forest for timber also worsens the situation. Against such a backdrop what we need is the sustainability of forest ecosystems, the proper valuation of the economic potential of this resource which goes beyond the amount of its carbon content along with the recognition of its contribution in supporting the livelihood of people.

1.4. *Loss of Biodiversity*

India is rich in biodiversity also. It has a great variety of plant species which amounts over 45,000 and over 90,000 species of animals.[55] The Northeast and the Ghats of the west coast contain large diversity of species in a relatively smaller geographical space. But India is suffering from extreme biodiversity loss today. Over 90 per cent of the area under the biodiversity hotspots in India has been lost according to the recent CSE report. Its wild flora and fauna are highly susceptible to climate change. An estimate shows that 10 per cent of the recorded flora is at risk.[56] Biodiversity loss emanates here from climate change, overuse of resources and desertification. Reducing genetic diversity and rising presence of alien species speed up the process too. Not only that India is a home to many medicinal plants which are also threatened due to the destruction of natural forests, woodlands and wetlands. Such habitat loss may contribute to depleting biodiversity that in turn may lead to low adaptability to environmental change.[57]

The climate change has exacerbated the problem as climate variability influences the durability of many wild animals and plants. Length of the day, precipitation pattern and temperature fluctuation control the life cycles of these species.[58] Wrong waste disposal and overuse of chemical pesticides are also creating problems.[59] A mega-diverse country like India can get leverage in the global market for genetic resources in agriculture and medicines as the Northern countries are poor in resources. But human greed and need along with climate change have destroyed these natural resources affecting both the economy and the ecosystem of the country. Before the Green Revolution and introduction of HYV seeds, it is reported that in India alone there were 50,000 varieties of rice. Now on an average, HYV of rice covers 70 per cent of cropland and 90 per cent is covered by wheat.[60] Such grim situation gives alert that India is on the brink of massive loss of biodiversity that will affect its economy and well-being.

1.5. Problems Pertaining to Energy Resources

Fossil fuel dependence in order to generate energy today gives birth to a kind of trepidation that we are heading towards our own ruin. As their production often entails huge environmental cost, there is a trend to replace fossil fuels with cleaner energy resources or to seek alternatives to conventional energy resources. Against such a scenario, India needs to address its energy challenges in congruence with environmental concern.

India is at an early stage of development. If she wants to keep the wheels of economic growth moving, it is impossible for her to stop fossil fuel consumption. In order to maintain her goal of 8–9 per cent economic growth, it would require about 1.5 tonnes of oil equivalent a year per person. With an expected population of 1.5 billion by 2025, the total energy requirement would be 2.25 billion oil equivalents a year. India's energy sector depends hugely on coal, while the power and industrial sectors require natural gas and oil is the moving force behind the transport sector.[61] (Figure 2.1) But India does not have complete energy independence. Therefore, she has to depend on energy-rich countries for import that again opens avenues of economic and political exploitation and affects her energy security. India has little reserves of oil, gas and Uranium but has large thorium reserve. India has only 5 billion barrels of oil reserves. Although new exploration licensing policy helps increasing production from domestic fields, there has been a consistent rise in oil imports. It has been estimated that most of the production depends on fields which are near saturation. India in the early part of 2020s might become the world's third largest among the net oil importing countries, following United States and China.[62]

Coal accounts for three-quarters of India's hydrocarbon reserves. It is abundant but due to its regional concentration, low content of ash, and small hydro-potential in comparison to our needs, its use for energy is relatively less.[63] The hydro-resources of the country are not only limited but

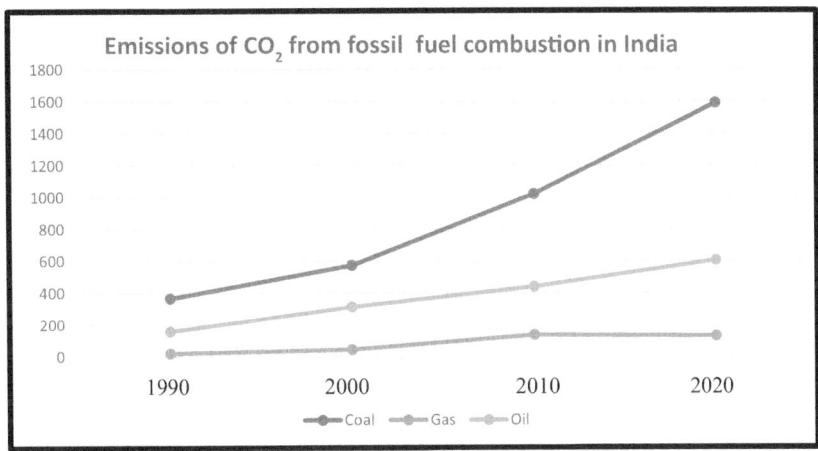

Figure 2.1 Emissions of CO_2 From Fossil Fuel Combustion in India

Source: Prepared by the author collating data from Ian Tiseo, *Carbon dioxide emissions from fossil fuel combustion in India from 1960 to 2020, by fuel type*, Statista, January 18, 2022. Available at India: fossil fuel emissions by type 1960–2020 | Statista Accessed on January 16, 2021.

also located in difficult regions of the Northeast and the Himalayan ranges. These projects have large financial costs, significant socio-environmental impacts and face a high hydrological risk due to receding glaciers and global warming led climate variability. Hydropower is expected to add less than 10 GW by 2020, which is less than 5 per cent of the incremental electricity needs by 2020 which is a marginal contribution.[64] While considering the potential of natural gas to facilitate a transition to low-carbon future for India, it is found that conventional gas reserves in India are limited. Most of our gas reserves are concentrated offshore, and 49 per cent of the gas fields are fast depleting.[65] This might lead to energy insecurity.

Given such scenario, India is today facing a twofold problem. First, it has to depend on imported oil and gas from the 'arc of energy'[66] which depends on foreign and diplomatic policy concerns. Although there is an abundant supply of indigenous coal and there exists high demand for this resource due to the high dependence of India's power sector on thermal energy, commercial inefficiency and 'infrastructural bottlenecks' may force her to import large quantities of coal in the foreseeable future. Coal imports have grown strongly over the past two decades.[67] The question of energy security has thus been cropped up. The result is that, first, energy supply is exposed to international markets on all fronts hampering her energy independence. Second, the problem associated with global warming as energy security is closely linked to it.

Fossil fuels are still the mainstay of India's energy mix. In order to ensure the adequacy of this resource, international cooperation is required. Increased financing is also required in a manner that may provide our population with clean energy at affordable cost. While climate change directly affects the renewable resources like food, water and forests, the pattern of using energy resources has a himalayan responsibility in polluting the environment. The associated human activities act only as a catalyst that creates catastrophes.

2. Extreme Climate Conditions in India

In India, climate-induced extreme weather events destroy lives, livelihood and property every year. Along with such disasters, rising sea level has also affected the country by inundating low-lying areas and offshore islands of the coastal belt. SLR affects the mangroves which are significant for preventing cyclonic hits to the coastal area. All of these adversely impact the country's socio-economic structures to a large extent.

2.1. Sea Level Rise and Cyclones

Accelerated sea level affects the Indian coast severely. It has led to the erosion of this region, and saline water ingression is a very common phenomenon. According to the Climate Risk Index, 2018, India was ranked fifth in the world on the basis of vulnerability to the climate change.[68] It is estimated that the melting of glaciers has contributed to about 45 millimetre (mm) of sea level rise during the past hundred years. Another cause for rising sea level is that the Himalayan ecosystem is highly vulnerable to climate change where glaciers have retreated more than 10 metres per year. Such retreat and ice melting from land glaciers may lead to further increase in sea level.[69]

India's long coastline and the entire coastal zone are fragile in the face of both slow- and rapid-onset natural calamities, which are extremely vulnerable to climate change induced fast- and slow-onset disasters. The ecosystem changes and saltwater intrusion due to sea level rise contaminates the groundwater, thereby affecting both the agricultural lands and the hydrological cycle of the country. Rising sea levels also result in the heightened frequency of coastal flooding due to storm surge events. Dramatic retreat of sandy beaches and beach erosion removing vital protective coastal features such as sand dunes and vegetation also exacerbate the risk of coastal flooding. Various studies have estimated the loss of lands and have predicted the vulnerability of forests on the coast due to SLR. An estimation projects the vulnerability of Indian coast due to 1 m of SLR for which nearly 14,000 sq. km coastal inundation would follow.[70] Another report enumerates the most vulnerable stretches of the Indian coast that include Mumbai, portion of Konkon coast, Kutch and Khambat and southern Kerala. The deltaic areas of Godavari, Krishna, Cauvery and Ganga are also at risk.[71]

The Sundarbans Island system is also threatened. An estimation of 2010 shows that sea level would rise closely 8 mm, which was a jump from 3.14 mm annual rise as recorded in 2000.[72] In a previous study by Professor Sugata Hazra and Anzi Baxi, it was revealed that some islands of the region would experience complete submergence like Kapasgadi and Lohachara. Lohachara was inundated in 1996. The report also said that the partial disappearance would result in some other islands like Sagar and Ghoramara.[73] Coastal erosion and submergence have been accentuated in the open coastal segment of Digha-Junput area over the last decade too. Sea wall had been built to protect the coastal area in Shankarpur and Mondarmoni – the tourist hubs. But the fisheries in Shankarpur were affected by SLR badly.[74]

The state of Gujarat which has the longest coastline (stretching across 1,700 km) among the Indian states is also vulnerable to accelerated sea level. Even 1-m sea level rise here may cause maximum damage to around 6 per cent of Gujarat's costal population. It is predicted that 0.181 million hectares area is likely to be inundated by the rising sea level as the coastal region is characterized by creeks and inland water – submergence type. Trend of accelerated sea level for Kandla port is 3.37 mm per year.[75] The popular beaches like Tithal and Dumas witnessed acute rise in sea level resulting in loss in tourism industry, which is adversely affecting coastal development and tourism industry. The vulnerability of Tamil Nadu is not less. Its coastal stretch is more than 1,000 km which have been destructed by the SLR.[76]

Local factors like subsidence also contribute to sea level rise. In Mumbai, sea level has increased by 0.77 mm/year as noted in past 100 years, while in Kolkata the rise was 5.22 mm/year during the past 50 years.[77] Both Mumbai and Kolkata cities are suffering similarly because of their low elevation.[78]

Apart from the sea level rise, intense storms accompanied by severe surges, embankment failure and recurrent coastal flooding may also result in adverse impacts on coastal wetlands and ecosystems. So far as India's coastline is concerned, the east coast is hit by recurrent storm surges every year and due to tropical cyclones generated in the Bay of Bengal, particularly vulnerable to the occurrence of storm surges generated by tropical cyclones in the Bay of Bengal. (Figure 2.2)Some of the devastating cyclones that affected the coast in the period 2007–2009 were Cyclone Sidr, Bijli, Aila, etc. They attacked the northern part of the Bay of Bengal. The 1990s experienced major cyclonic storms as well that ravaged the Indian Coastal area. In 1996, the worst cyclone of the century killed 2000 residents in Andhra Pradesh. The Supercyclone of Orissa of October 1999 was also disastrous.[79] In the year 2013, another devastating super cyclone Phailin hit the Orissa coast severely and in 2014, cyclone Hudhud caused havoc to Andhra Pradesh. The recent Cyclone Fani, Amphan and Yash were also devastating, especially Amphan that hit West Bengal dangerously in 2020 amidst the COVID-19 pandemic caused huge economic loss.

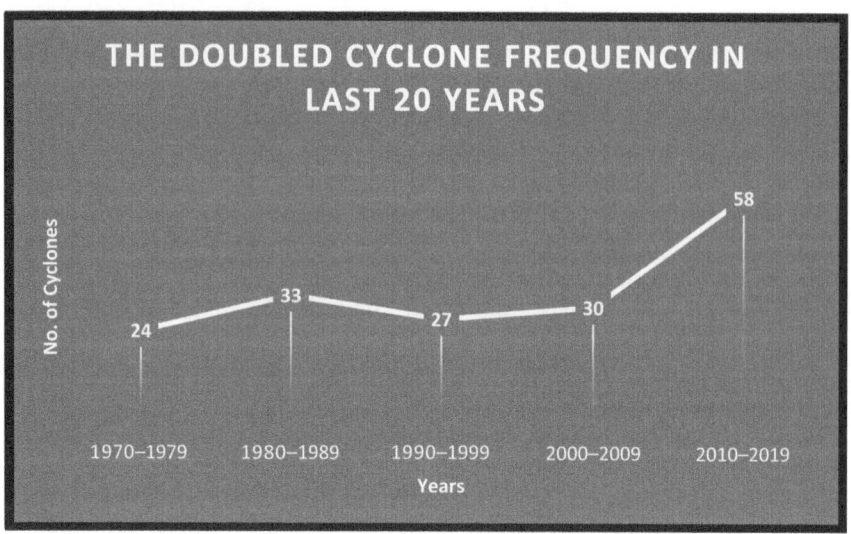

Figure 2.2 The Doubled Cyclone Frequency in Last 20 Years

Source: Rai, Dipu and Mishra, Mayank, "Rising frequency and intensity of flood, drought and cyclone", India Today, June 12, 2021, available at Rising frequency and intensity of flood, drought and cyclone – DIU News (indiatoday.in) Accessed on December 19, 2021.

West Bengal also witnessed several such 'rapid-onset' type disasters. Here, in the last 2 decades large amount of crop and forest property had been destroyed or lost and over millions of people had been affected. The Indian Sundarbans Delta is severely affected where more than 1 million people are at risk of SLR.[80] Storm surges have devastated the area near West Bengal also, as happened during the 2009 Aila.

The Tamil Nadu coast has undergone hindrance to the natural process of sediment movement to northward and southward due to severe development pressure on the coast along with erosion and accretion. Such obstruction to natural process has altered the coastline rendering it fragile to storm surges, cyclones and consequent inundation of low-lying areas. It also witnessed frequent cyclonic storms. The 2004 tsunami demonstrated the vulnerability that the coast faces as many natural ecosystems including the mangroves and fresh water reservoirs were damaged.[81] The Andamans, particularly Car Nicobar island, and Tamil Nadu, are the worst hit, although other States such as Kerala and Andhra Pradesh have been affected by the killer waves in varying degrees.[82] The worst affected Nagapattinam village showed longer penetration of sea water up to an elevation of 3.9 m due to the gentle slope of coastal land combined with the effect of tsunami wave.[83]

While the frequency, duration and intensity of tropical cyclones and storms are causing unmanageable havoc in the coastal belts of India, large urban centres are not free from fear of devastation. Cyclone and storm surges could have impact on the megacities also among which Mumbai, Chennai and Surat are significant.[84] Although it is true that global warming helps largely in altering coastal morphology and coastal characteristics by sea level rise induced coastal inundation and with rising intensity of cyclones and storm surges, unrelenting human activities along the coast including port building, groundwater extraction, aquaculture, agriculture as well as rising human settlements in the coastal area may escalate the fragility of coastal region of India. All of these lead to shoreline retreat even coastal flooding in cities as mentioned earlier. The ultimate result is financial loss and loss of lives and property.

2.2. *Increased Intensity of Natural Disasters*

Along with the vagaries of climate change as experienced by the coastal region of the country, the frequency and magnitude of extreme weather events have also affected other parts of the country. The IPCC fifth assessment report also warned that the intensity of rainfall in India would increase due to climate change. It has been estimated that the monsoon floods hit over 30 million people in India where ironically 60 per cent of farmland and livestock and the forest area need sufficient rains for survival.[85]

In the Himalayan ecosystem, global warming and resultant climate change are gradually transforming the ecological and socio-economic landscape. This seismic zone is prone to earthquakes and flashfloods creating pressure on lives and economy. Human activities and unplanned development works and spread of tourism have played a critical role in augmenting such disasters. There are several instances of various climate catastrophes in the Himalayan region like the Alakananda disaster (1970), where a major landslide blocked the Alakananda River and inundated thousand kilometres of land, which washed away numerous bridges and roads and Tawaghat tragedy (1978) which took an even greater toll when an entire mountain slope collapsed into the Bhagirathi River, forming a lake that burst and flooded the Gangetic plain. Such events were a wakeup call to the government about the ill-effect of unscrupulous destruction of forest catchments that may result in disasters. These wakeup calls remained unattended to.[86] Extreme deforestation also results in devastating flood in Assam. In July 2012, the Brahmaputra and its tributaries showed its wrath by flooding the Kaziranga National Park that forced the animals to move to higher ground. The Island Majuli was the worst affected as it was entirely inundated leaving 75 families homeless.[87] Erosion of soil and recurring floods are very common in the Northeast region too. During the period of 45 years (1953–2004) due to floods the region suffered a financial loss of over 1 crore with loss in agriculture and million hectares of land. Assam, Meghalaya, Manipur, Tripura and Nagaland were affected by soil erosion too.[88]

Bursting of glacial lakes is a common phenomenon in the hills, so is the case in India as she has been devastated several times by such glacial lake burst induced flashfloods. Human activities which are unmindful of the geo-ecological fragility have worsened the situation. In 2004, a glacial lake that had formed along the course of the Tsang Po in China had burst leading to a flash flood across the border in Arunachal. The effects were tremendous as unscrupulous urbanization, industrialization and development plans have made the region more susceptible to such catastrophes. The building of a series of dams in Uttarakhand and other parts of the Himalayan Region has similarly destroyed the ecological cycle of the region. Uttarakhand – the abode of the gods – has paid the price for that by witnessing the most devastating flood of the decade in June 2013. Flash floods resulting from extremely intense rainfall on 16 June 2013 devastated the state of Uttarakhand. The scientific community while analysing the disaster said that the changing climate of the earth to a large extent is responsible for the disaster.[89] Uttarakhand, Himachal Pradesh and Northern part of West Bengal have experienced extreme devastation due to landslides and torrential rains in October 2021 similarly. In the last 20 years, such climate cataclysms are very common. In 2003, Tehri also witnessed such torrential rainfall, as occurred in 2010 in Ladakh leading to loss of property and lives.[90] In September 2014, the Jammu and Kashmir region was also terribly affected by devastating flood which was the result of intense and unprecedented rain.

Climate change might trigger the disasters, but the catastrophe is human-induced. In case of Uttarakhand disaster of 2013, the impact of cloud-burst would not have been so high if the state had taken a more regulated approach regarding the unscrupulous development works along the mountain slopes. Other Himalayan states are not beyond danger. The respective state government policies regarding construction activities in the region must be therefore subject to scrutiny. Huge rains have also wreaked the socio-economic structure of the Western part of the country. In June and July 2005, the country witnessed two devastating floods due to torrential rain in the states of Gujarat and Maharashtra which hit the economy hard. Mention must be made here about the 2005 flood in Mumbai that caused more than $1 billion financial loss as estimated.[91] These events proved the IPCC 4th Assessment Report prediction that extreme rainfall events are set to increase over the Indian subcontinent.

Along with the Northern and Western part of India, eastern fringe of the country specifically Uttar Pradesh, Bihar and Assam is bearing the brunt of floods. Often human activities are largely responsible for such disasters. The opening of lock gates of dams and barrages causes floods in some of these states. Breach of embankment also created havoc as happened in 2008 at the embankments of Kosi River that affected several districts of Bihar and Nepal. The Kosi flood dismantled the fact that there existed large failure in structural approach to flood control and institutional dysfunction with respect to transboundary flood management. It became evident that the preparedness for disasters and awareness among common masses and absence of priority and urgency in maintaining the flood control structures contributed to the embankment breach in 2008.[92] Man-made floods have also taken place due to release of water from barrages even during the time of rains. The people of Krishna and Guntur districts of Andhra Pradesh experienced this from time to time.[93]

All the severe weather events have created havoc in the country. The catastrophes created by such natural calamities are human-induced to a great extent. All of these incidents uproot people from their native place creating thousands of both internally displaced persons and climate migrants crossing the borders. So management of disasters both natural and man-made is absolutely essential in a country like India in order to contain and lessen deaths and damages. The National Disaster Management Board should act more efficiently to counter the wrath of nature and the resultant calamities, and the government is also supposed to give clearance to only those projects having minimum environmental and social cost as we need to concentrate on ways of development without destroying natural resources.

3. Environmental Change Induced Displacements in India

Displacements due to environmental hazards in India are very common and inevitable. It has been estimated that in 2050, climate disasters would generate more than 4 crores of climate migrants across India. According to the report, the average number of climate migrants in India during the period 2011–2020 is 33,78,000 which is higher than conflict-driven displacements.[94] Climate change triggers several types of displacement in India. First, natural disasters that include droughts, floods, volcanoes and earthquakes may induce displacements of people. Second, slow-onset changes in climate like desertification, SLR, degradation of land and forests induced displacement. Third, industrial accident led temporary dislocation and, finally, development project and urbanization induced climate refugees. In addition to this, climate change might lead to an influx of migrants from countries in the surrounding regions who may strain the resource base of the country as well.

3.1. Drought-Induced Displacements

Drought impacts the lives and livelihood of people gravely. As low farm output and land degradation are inevitable condition of drought, people do not find any other options but to seek sanctuary in other locations. Every year thousands of people are migrated from the drought-prone Bolangir district of western Orissa. The western part of India is severely drought-prone and climate change has exacerbated the situation. In this part of India, Luni River is susceptible to acute physical water-scarce conditions.[95] Similar condition exists in basins of Mahi, Pennar, Sabarmati and Tapi Rivers. In the large area of Western and Southern parts of the country, increased drought situation affects subsistence agriculture. As a result, much of the poor marginal peasants are forced to relocate to cities. Rural-to-urban migration for drought-induced uninhabitability is very common in India.[96]

3.2. Extreme Weather Events Driven Displacements

Increased and erratic rainfall induced floods in India as has already been discussed in the previous section may cause huge damages to livelihood resources and life support systems, thereby enhancing mass migration as well. Among the most flood ravaged countries of the world, India ranked second after Bangladesh. Annually 30 million people are uprooted for this.[97] Recurrent monsoon floods are very common here. The state of Bihar is frequently affected by such floods and sometimes breaching of embankments further the process at the detriments of lives, livelihoods and property. During the 2008 Kosi flood due to breach of embankment, over three million inhabitants were rendered homeless and it was considered the worst flood in the state during the last 50 years. Flood-related deaths, diseases and injuries have impacted the social fabric of the affected areas in Bihar due to the calamity.[98] In 2012, the recurring and huge floods had ravaged the northeastern states of Arunachal Pradesh and Assam. The displacement that followed was the largest as more than 6 million people were displaced.[99] Another devastating flash flood in Uttarakhand in the same year has generated such temporary displacements as near about 277 villages have been ruined by the catastrophe and survivors have become refugees on their own land.

Apart from floods and heavy rains, cyclones, another rapid-onset climate disaster, create dislocation and displacements. In the year 2009, the severe cyclone 'Aila' hit the state of West Bengal. The intensity of the cyclone was deadly as the speed of it was 120 km/hr which was associated with severe storm surge.[100] Since most villagers usually depend on agriculture in the cyclone-affected Kumirmari island, Lahiripur, Char Bidya and Masjidbari, the uncultivable land left lakhs of people jobless. Thousands of such storm-battered, hungry refugees from the Sundarbans fled to Kolkata in search of jobs and refuge.[101] Another cyclone, namely, Laila again caused havoc as heavy rains accompanied by high-speed winds battered six coastal districts of Andhra Pradesh in 2010. At least 50,000 people living in low-lying areas and isolated islands on the Godavari and Krishna deltas were evacuated and relocated to safer places.[102] Cyclone Amphan of 2020 and Yash of 2021 also caused havoc to West Bengal creating thousands of internally displaced persons.

All of these displacements and relocations of people due to such rapid-onset climate hazards are usually temporary in nature. After the catastrophes usually people of this kind return back to their native place, and due to their temporary migrations, they can be termed more relevantly as internally displaced persons rather than refugees. However, in sharp contrast to other causes that force people to seek sanctuary in other places, climate change and concomitant sea level rise generate permanent uprooting of affected people.

3.3. Sea Level Rise Induced Displacements

India is highly vulnerable to sea level rise, but a large share of its population reside in and around the low elevated coastal zone.[103] An estimation shows that the projected 1 m sea level rise might result in the loss of huge land and forests in India leading to displacement of 7.1 million people approximately.[104]

The Sundarbans island system is the most vulnerable. The vanishing islands of this delta give birth to many climate refugees. In this delta, some islands are vanishing from the map rendering thousands of people permanently homeless. Numerous people are dislocated from their native places from the islands like Ghoramara, Sagar, Bhangaduni and Dalhousi as these islands are facing erosion and submergence. During the last several years, the Sagar Island alone lost an area of 30 km². Even people from the villages of Khasimara, Baghpara Island, etc. sought relocation to Sagar Island also.[105] In 1996, displaced people from the island Lohachara were also forced to flee as the island disappeared altogether. Along with sea level rise, recurrent flooding, severe cyclonic storms have devastated the area too, which uproot around 600 families from islands like Ghoramara, Mousuni and G-Plot, and transformed them into climate refugees.[106] Megacities are not escaped such as Mumbai and Kolkata where such displacements are also taking place.[107]

While in India environmental degradation has played a crucial role to instigate popular displacement, the country is also receiving continued exodus of climate-induced migrants from Bangladesh. The reasons behind the relocations in India from Bangladesh are not always political religious or ethnic. In Bangladesh, limited resource base often pushes people to migrate. Natural resource scarcity triggered by environmental change has further been worsened by unequal accessibility to these resources. Along with these recurring cyclones, riverbank erosion, floods and sea level rise also forced people to move to the neighbouring country for shelter. Over 120 million people might become migrants because of sea level rise in both India and Bangladesh.[108] Many people from Bangladesh crossed border and relocate to India for flood-induced problems also. Development projects including hydroelectric ones also give birth to refugees crossing border to seek refuge in India. Mention may be made here about the Kaptai Dam issue and migration of Chakmas from Bangladesh to the northeastern part of India.[109] River linking projects by India have also raised eyebrows in Bangladesh, as being a low riparian to India it is fearing displacements due to the construction of reservoir.[110]

3.4. Development-Induced Displacements Causing Environmental and Social Harms

Development-induced displacement is a matter of serious concern for India too where industrial projects, dams, roads, mines and power projects have not only grabbed lands but also displaced millions of people creating tremendous pressure on society and environment. Particularly, dams create controversy in India as they often displace local people. It has often been criticized severely, and there exists a disenchantment with large projects over more than past two decades. The environmental hazards and displacements induced by such projects are at the heart of controversy in the country as the project construction in remote and often pristine areas usually disturbs natural habitats of the tribal community.[111]

The Silent valley project in Kerala, the Pong dam in the northern state of Himachal Pradesh, the Narmada project in Gujarat and the Tehri Hydro Electric Project in the Himalayan Region have been facing strong anti-project movements due to both the distortion of ecological fabric of the regions and the resultant internal displacement. The first successful transnational campaign opposing a big dam project in India was waged against the Silent Valley Project in the 1970s and 1980s. Around the same time, initial domestic resistance to the building of the gigantic Tehri dam at the foot of the Himalayas had begun a struggle that under various incarnations continued into the 1990s. These movements especially the Silent Valley victory inspired big dam opponents throughout India. The momentum carried through to a campaign against a series of hydro-projects planned for the Godavari and Indravati rivers in Central India. The Bodhghat, Inchampalli and Bhopalpatnam dams also met with protests as they were to dislocate over 110,000 tribals and flood many tens of thousands of hectares of forests, including part of the Indravati Tiger Preserve Sanctuary habitat of the Indian wild buffalo.[112] The most well-known Narmada Bachao Andolan is also against dam-induced displacement as the Sardar Sarobar Project uprooted hundreds of people. It was claimed by the Madhya Pradesh government that adequate compensation had been disbursed among the 45,000 displaced persons.[113] But the victims were not satisfied with the government activities regarding resettlement and rehabilitation.

The hilly landscape of Northeast India is suffering from massive distortion of the ecology due to the open cast mining and construction of hydropower projects. The Gumti dam of Tripura, Loktak hydroelectric project of Manipur displaced many people. The 105 MW Loktak hydroelectric power project proponents did not take into account the effects on the lake ecosystem, the people and wildlife whose lives are connected with the lake in their cost–benefit analysis. Within a few years of the project's commissioning, it became evident that the lake was undergoing drastic changes.[114] The

Pagladiya project of Assam and Tipaimukh project of Manipur faced hard opposition because of the fear of displacement also.[115]

All of these events have proved that how massive man-made distortion of the environmental fabric can induce insecurity. Along with dam building, other development activities as desired by the successive governments of India like mining and building of industrial infrastructure by the private conglomerates and multinationals have also encroached upon rural lands ousting and uprooting thousands of people from their traditional habitats.[116] Most of these projects have violated the environmental norms of the country. Like the Vedanta Project of Orissa, Mining Corporation on the Niyamgiri Hills violated the landmark environmental legislations of India. However, as the project failed environmental test, it stood cancelled. Dongria Kondh tribals of Niyamgiri Hills in Odisha had played a major role in stalling this project. They, supported by the environmentalists, have been strongly opposing the mining leases on the ground as the projects were environmentally disastrous to the local tribes. This kind of decision was the first acknowledgement by the government that the aforesaid laws bind them.[117] Without full consultation with the predominantly Indigenous communities before any efforts of natural resource development in the pristine areas thus exacerbates tensions in India. The South Korean steel giant POSCO project was another one of many controversial development plans throughout the country. The project has also created mass displacement, which has faced the denial of their forest rights resulting in the loss of the entire social texture into which they were born. The Jindal Buxite Mining project, Careamol Iron Ore Mining, Goa, Nagarjuna Construction Limited Power Project, Andhra Pradesh, the Tata Nano project in West Bengal, have also been protested as the locals alleged that either the Forest Right Act has not been implemented and rights of traditional forest dwellers were not met by the project planners and the government or the projects were proposed in ecologically sensitive areas or there exist illegal and forced land acquisition hampering among the ecological, economic and food security of the original landowners. However, sometimes opposition to such development-induced displacement proved to be a dangerous undertaking. For instance, an environmentalist, namely, Shehla Masood, was killed in August 2011, for campaigning against the mining company Rio Tinto in Madhya Pradesh.[118] So, development affects the real custodian of nature, that is, the Indigenous population the most. They are the traditional agents of natural conservation too. They view their severance from an ecosystem which had sustained them, as an existential threat.

4. Summary

Global warming may thus have severe consequences for India. The scarcity of resources, coastal inundation and severe weather events all are risks that the people of the country are facing in reality due to climate change. In this chapter, the various kinds of resource scarcity that are triggered by climate change have been discussed. Poverty and human-constructed issues have exacerbated those problems. The poorest sections of India which are suffering from the greatest levels of chronic hunger and malnourishment are exposed to both the unavailability of food production and the instability of food supplies. The situation gets worsened while the affordability of buying food products is minimized and allocation of food products in accordance with each person's entitlement to a particular food product is hindered or undermined. In case of climate-induced drops in production and consequent food scarcity, the national interest of India can be affected in two ways. It may directly hit the economy, as 30 per cent of India's gross domestic products depends on agriculture.[119]

The following diagram (Figure 2.3) compares two scenarios where the economic condition of the same country differs in two climatic situations:

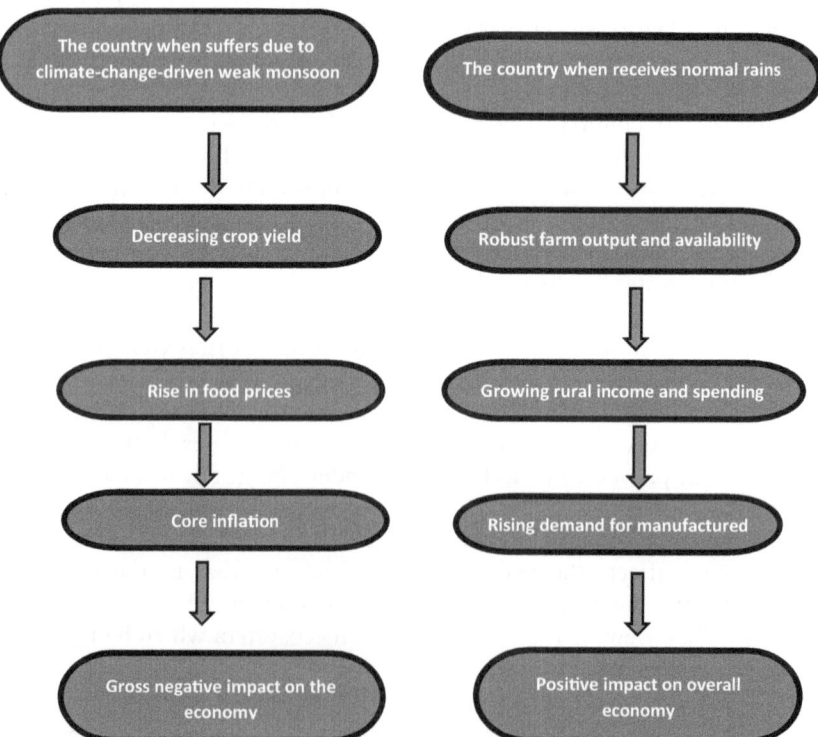

Figure 2.3 Economic Conditions of a Country Under Two Scenarios

Source: Prepared by the author.

This kind of food scarcity due to climate change affects the economic well-being of our country in the same manner. Second, jostled demands may arise on these diminishing resources from which inter-community and inter-class violent conflict may ensue. There are different reasons behind hike in food prices due to its scarcity and the increase in number of malnourished people in the country. But environmental change which affects food production to a great extent, though may not be an igniter of political tussle over this event, often multiplies or amplifies the propensity of the problem.

Regarding climate change induced scarcity of water, human-constructed issues have also responsible for declining water table. Due to lack of sufficient water for intensive agricultural activities which are necessary to feed the burgeoning population, unsustainable irrigation activities take place affecting the resource. Additionally, to meet the demand of rising population, over-pumping and extraction of the groundwater are very common phenomena here. The country has also been plagued with the problems of inadequate water storage capacity and water management policy. Subsidized water service in some places furthered the problem as it may result in overconsumption and wastage. Moreover, water scarcity also intensifies the competition between municipal, industrial and agricultural demands. The powerful sectors like industry may capture more water through buying and bullying major sources at the expense of subsistence agriculture of the rural poor. Market intervention has exacerbated the problem by making the resource a commodity.

It is unquestionable that the poor and marginalized of the society are the hardest hit as they are dependent on natural resources more than the urban rich and middle-income groups. Environmental change directly affects these primary-resource-dependent groups since their natural support system is at stake due in part to overuse and unsustainable consumption and partly to global warming and climate change driven scarcity. The weak governance system then acts as only a catalyst in the process of problem escalation.

Presently, thus India is confronted with several challenges regarding the environmental change issue. There exists a dilemma that the country is facing hugely – the dilemma between environmental protection and economic growth. This issue basically highlights the fault line between the need for ecological preservation for a country like India and her aspiration for development. As an emerging global power, India while does not want any disruption in her development prospects for securing her goal of achieving economic growth, the present global scenario also demands her to take a leadership role in the ongoing climate bargains. Any sluggish response on her part to the mitigation and adaptation process to the problem may act to her own detriments.

Notes

1 Catherine Folley and Andrew Holland, *Climate Security Report-Part Two, Climate Change and Global Security*, October 2012. Available at http://americansecurityproject.org/reports/2012/part-two-climate-change-global-security/ Accessed on May 17, 2013.

2 The Energy and Resources Institute (TERI), *Looking Back to Change Track: Green India 2047*, p.2. Available at www.Icsudev.org/summary_greenindia.pdf Accessed January 5, 2013.

3 Praful Bidwai, *The Politics of Climate Change: Mortgaging Our Future*, Orient Blackswan: Hyderabad, p.52.

4 Peter H. Gleick, "Climate Change and International Politics: Problem Facing Developing Countries", *Ambio*, Vol.18, No.6, 1989, p.337.

5 Zia Haq and Gautam Choudhury, "India's Mid-Summer Nightmare", *The Hindustan Times*, July 29, 2012.

6 P.K. Gautam, "Climate Change and Conflict in South Asia", *Strategic Analysis*, Vol.36, No.1, January, 2012, pp.32–40.

7 "Desertification Affects Over One-Fourth Area of India: Report", *The Hindustan Times*, April 28, 2014.

8 Akash Vashishtha, *Desertification Affects Quarter of India's Land*, June 18, 2014. Available at indiatoday.intoday.in/story/desertification-affects...indias.../367345.html Accessed on November 18, 2014.

9 Indian Network for Climate Change Assessment (INCCA), *India: Greenhouse Gas Emissions 2007*, May 2010, p.22. Available at http://www.moef.nic.in/downloads/public-information/Report_INCCA.pdf Accessed on December 14, 2015.

10 Pandurang Hegde, "Need to Preserve Our Bio Diversity", *Deccan Herald*, October 10, 2012.

11 Satyarat Chaturvedi, "GM Crops Are No Way Forward", *The Hindu*, August 24, 2012.

12 "Diabolic Game", *Deccan Herald*, August 13, 2012.

13 Ministry of Water Resources, Government of India, *National Water Mission under National Action Plan on Climate Change: Comprehensive Mission Document*, Volume – I, April 2009, pp.1–2. Available at www.nicra-icar.in/.../Mission%20Documents/WATER%20MISSION.pdf Accessed on June 12, 2013.

14 "Asia Faces Risks of Disease, Hunger, Document: Working Group-II Contribution to the IPCC 4th Assessment Report", *The Asian Age*, April 7, 2007.

15 IPCC, *Analysing Regional Aspects of Climate Change and Water Resources*, p.80. Available at www.ipcc.ch/pdf/technical-papers/ccw/chapter5.pdf Accessed on March 20, 2015.

16 Ashok Jaitly, "South Asian Perspectives on Climate Change and Water Policy", in David Michel and Amit Pandya (eds.), *Troubled Waters: Climate Change, Hydropolitics, and Transboundary Resources*, The Henry L. Stimson Center: Washington, DC, 2009, p.20. Available at www.stimson.org/images/uploads/.../Troubled_Waters-Complete.pdf Accessed on June 13, 2013.

17 Holger Treidel, Jose Luis, Martin-Bordes and Jason J. Gurdak (eds.), *Climate Change Effects on Groundwater Resources: A Global Synthesis of Findings and Recommendations*, Taylor & Francis Group: London, 2012, p.2.

18 Himanshu Kulkarni and Himanshu Thakkar, "'Framework for India's Water Resource Management Under a Changing Climate", in Navroz K. Dubash (ed.), *Handbook of Climate Change and India: Development, Politics, and Governance*, Oxford University Press: New Delhi, 2012, p.332.

19 "India Stares at Uneven Monsoons", *The Times of India*, June 20, 2013.

20 "Averting Water Crisis", *The Hindu*, July 9, 2007.

21 World Bank, Agriculture and Rural Development Unit, South Asian Region, Report No. 34750-IN, *India's Water Economy: Bracing for a Turbulent Future*,

December 22, 2005, p. XI. Available at www-wds.worldbank.org/external/default/.../WDSP/IB/.../34750.pdf Accessed on May 29, 2013.

22 "For Land and Water", *Economic and Political Weekly*, Vol.XLVI, No.34, August 20, 2011, p.9.

23 Vandana Shiva, *Water Wars: Privatization, Pollution and Profit*, South End Press: Cambridge, MA, 2002, pp.107–110.

24 Ibid., pp.110–111.

25 Vandana Shiva and Vinod Kumar Bhatt (eds.), *Climate Change at the Third Pole: The Impact of Climate Instability on Himalayan Ecosystems and Himalayan Communities*, Navdanya/Research Foundation for Science, Technology and Ecology: New Delhi, 2009, p.13.

26 "Rivers in Danger", *The Hindu*, March 24, 2007.

27 Praful Bidwai, *The Politics of Climate Change and the Global Crises: Mortgaging Our Future*, Orient Blackswan: New Delhi, 2012, pp.56–57.

28 Mats Eriksson et al., *The Changing Himalayas: Impact of Climate Change on Water Resources and Livelihoods in the Greater Himalayas (ICIMOD)*, December 11, 2009, pp.4–7. Available at www.worldwatercouncil.org/...water.../climate_change/PersPap.01._The... Accessed on June 25, 2013.

29 Ramaswamy R. Iyer, *Water: Perspectives, Issues, Concerns*, SAGE: New Delhi, 2003, p.259.

30 Brahma Chellaney and Heela Najibullah, *On the Frontline of Climate Change: International Security Implications*, KAS Publication Series No.17: New Delhi, 2007, p.37.

31 Vandana Shiva and Vinod Kumar Bhatt (eds.), No.30, p.14.

32 Ramaswamy R. Iyer, No.29, p.222.

33 In fact, China has been damming most international rivers flowing out of Tibet, whose fragile ecosystem is already threatened by global warming. The only rivers on which no hydro-engineering works have been undertaken so far are the Indus, whose basin falls mostly in India and Pakistan, and the Salween, which flows into Burma and Thailand (Chellaney 2009). So China has water disputes with various countries like in Mekong river Basin with Cambodia, Laos, Vietnam, Thailand and Burma due to damming of the river upstream, in Brahmaputra river basin with India and Bangladesh, in Sutlej River Basin with India and Pakistan, and so on. Brahma Chellaney, *On the Frontline of Climate Change*, Konrad-Adenauer-Stiftung: New Delhi, 2007, pp.59–60.

34 Chellaney Brahma, "Climate Risks to Indian National Security", in David Michel and Amit Pandya (eds.), *Indian Climate Policy: Choices and Challenges*, The Henry L. Stimson Center: Washington, DC, November 2009, p.26. Available at www.stimson.org/images/uploads/research-pdfs/fullreport.pdf Accessed on June 26, 2013.

35 Krishnan Ananth, "China Reassures India on Dam Projects", *The Hindu*, November 6, 2009.

36 Mark Christopher, *Water Wars: The Brahmaputra River and Sino-Indian Relations, CIWAG Case study on Irregular Warfare and Armed Groups*, US Naval War College, 2013. Available at https://www.usnwc.edu/getattachment/b5236b30-fce4.../Water-Wars.pdf Accessed on July 14, 2014.

37 Chellaney Brahma, *Water: Asia's New Battleground*, Georgetown University Press: Washington, DC, 2011, p.149.

38 Centre for Education and Documentation, *Forests and Climate Change in India*. Available at http://base.d-p-h.info/en/fiches/dph/fiche-dph-8613.html Accessed on July 15, 2013.

39 N.H. Ravindranath, N.V. Joshi, R. Sukumar and A. Saxena, "Impact of Climate Change on Forests in India", *Current Science*, Vol.90, No.3, February 10, 2006, pp.354–361. Available at www.currentscience.ac.in/Volumes/90/03/0354.pdf Accessed on July 13, 2013.

40 Jayashree Nandi, "'Depleting Forest Cover Could Have Grave Impact", *The Times of India*, June 21, 2011.
41 M.K. Pandit, Navjot S. Sodhi, Lian Pinkoh, Arun Bhaskar and Barry W. Brook, "'Unreported Yet Massive Deforestation Driving Loss of Endemic Biodiversity in Indian Himalaya", *Biodiversity Conservation*, Vol.16, 2007, p.354.
42 Maharaj K. Pandit, "Nature Avenges Its Exploitation", *The Hindu*, June 21, 2013.
43 Journals of India, *India State of Forest Review*, January 15, 2022. Available at India State of Forest Report, 2021 – Journals of India Accessed on January 22, 2022.
44 N.G. Hegde, *Challenges of Community Forestry in India, Asia Pacific Forestry Research – Vision 2010*, Proceedings of the Regional Seminar, Kuala Lumpur, Malaysia, 2000, p.13. Available at www.fao.org/3/a-w7732e.pdf Accessed on December 21, 2014.
45 Vandana Shiva, *Water Wars: Privatization, Pollution and Profit*, South End Press: Cambridge, MA, 2002, pp.44–45.
46 Ibid.
47 Madhav Gadgil and Ramachandra Guha, "Ecological Conflicts and the Environmental Movement in India", *Development and Change*, Vol.25, 1994, p.104.
48 Jayashree Nandi, "World Environment Day 2013: Remembering Silent Valley Movement", *The Times of India*, June 5, 2013.
49 Manshi Asher, "Renuka Dam: The Saga Continues", *Economic and Political Weekly*, Vol.XLVII, No.32, August 11, 2012, pp.31–32.
50 Vandana Shiva, No.67, p.8.
51 Gethin Chamberlain, "Indian Forest Villagers Rise Up to Halt UK Firm's Bid to Clear Land for Mining", *The Guardian*, June 28, 2014. Available at http://www.theguardian.com/world/2014/jun/28/india-controversial-forest-coalmine-essar-energy Accessed on January 15, 2015.
52 Kumar Sambhav Shrivastava, "Green Tribunal Spells Its Mandate", *Down to Earth*, January 1–15, 2013, p.16.
53 "Helping Forests Disappear", *Economic and Political Weekly*, Vol.XLVII, No.8, February 23, 2013, p.8.
54 Smriti Das, "The Strange Valuation of Forests in India", *Economic and Political Weekly*, Vol.XIV, No.9, February 27, 2010, p.16.
55 Ministry of Environment and Forest, Government of India, *India's Fifth National Report to the Convention on Biological Diversity*, 2014. Available at https://www.cbd.int/doc/world/in/in-nr-05-en.pdf Accessed on November 14, 2014.
56 S. Balaji, "Biodiversity Challenges Ahead", *The Hindu*, May 27, 2010.
57 Kartikeya Sarabhai, *What Are the Threats to Biodiversity*, p.4. Available at http://www.vigyanprasar.gov.in/Radioserials/Threat_to_biodiversity_-_draft_paper.pdf Accessed on November 15, 2014.
58 Vinod Kumar and A.K. Chopra, "Impact of Climate Change on Biodiversity of India with Special Reference to Himalayan Region: An Overview", *Journal of Applied and Natural Science*, Vol.1, No.1, 2009, pp.117–122.
59 Ibid.
60 Col. P.K. Gautam, *Environmental Security: Internal and External Dimensions and Response, A United Service Institution of India Project*, DS Kothari Chair, Knowledge World: New Delhi, 2003, p.36.
61 Bhupendra Kumar Singh, India's Energy Security: Challenges and Opportunities", *Strategic Analysis*, Vol.34, No.6, November 2010, p.801.
62 International Energy Agency, *World Energy Outlook 2007: China and India Insights*, 2008, p.489. Available at http://www.worldenergyoutlook.org/media/weowebsite/2008-1994/weo_2007.pdf Accessed on December 30, 2014.

63 Planning Commission, Government of India, *Integrated Energy Policy*, Report of the Expert Committee, August 2006, pp.33–35. Available at http://planningcommission.nic.in/reports/genrep/intengpol.pdf Accessed on September 5, 2013.

64 Girish Sant and Aswin Gambhir, "Energy, Development and Climate Change," in Navroz K. Dubash (ed.), *Handbook of Climate Change and India: Development, Politics, and Governance*, Oxford University Press: New Delhi, 2012, p.295.

65 International Energy Agency, No.63, pp.510–515.

66 Stretching from the gulf region to the Caspian Sea through Siberia and the Arctic region to the Russian Far East, Alaska and Canada. Nearly 80 per cent of the world's oil and gas, including potential reserves, are located in this region. Asian countries will remain dependent on energy from this arc.

67 International Energy Agency, No.63, p.509.

68 Mohanty Abinash, "Preparing India for Exteme Climate Events", *Hindustan Times*, July 1, 2021. Available at https://www.hindustantimes.com/ht-insight/climate-change/preparing-india-for-extreme-climate-events-101625127345593.html Accessed on January, 19, 2021.

69 J. Srinivasan, "Impacts of Climate Change on India", in Navroz K. Dubash (ed.), *Handbook of Climate Change and India: Development, Politics, and Governance*, Oxford University Press: New Delhi, 2012, pp.33–34.

70 M. Zafar-ul Islam, Shaily Menon, Xingong Li and A. Townsend Peterson, "Forecasting Ecological Impacts of Sea-Level Rise on Coastal Conservation Areas in India", *Journal of Threatened Taxa*, Vol.5, No.9, May 26, 2013, pp.4349–4358. Available at http://dx.doi.org/10.11609/JoTT.o3163.4349-58 Accessed on July 24, 2013.

71 Ministry of Environment and Forests, Government of India, *India's Second National Communication to the United Nations Framework Convention on Climate Change*, 2012, pp.149–150. Available at http://www.moef.nic.in/downloads/public-information/India%20Second%20National%20Communication%20to%20UNFCCC%20Executive%20Summary.pdf Accessed on March 20, 2015.

72 "Rise in Sea Level Threatening Sundarbans", *The Hindu*, January 6, 2013. Available at http://www.thehindubusinessline.com/news/science/rise-in-sea-level-threatening-sundarbans-sayspachauri/article4279603.ece?ref=relatedNews Accessed on July 26, 2013.

73 Sugata Hazra and Anji Bakshi, "Environmental Refugees from Vanishing Islands", in Purusottam Bhattacharya and Sugata Hazra (eds.), *Environment and Human Security*, Lancer's Book: New Delhi, 2003, pp.220–221.

74 Sugata Hazra, *Climate Change Adaptation in Coastal Region of West Bengal*, Climate Change Policy Paper II, pp.10–13. Available at http://awsassets.wwfindia.org/downloads/climate_change_adaptation_in_coastal_region_of_west_bengal. pdf Accessed on July 26, 2013.

75 Rajiv K. Gupta, *Climate Change and Water Resources Management in Gujarat*, Climate Change Department, Government of Gujarat, pp.9–10. Available at www.aragon.es/estaticos/GobiernoAragon/.../Rajiv%20Gupta.pdf Accessed on July 28, 2013.

76 Sujata Byravan, Sudhir Chella Rajan and Rajesh Rangarajan, "Sea Level Rise: Impact on Major Infrastructure, Ecosystems and Land along the Tamil Nadu Coast", in Navroz K. Dubash (ed.), *Handbook of Climate Change and India: Development, Politics, and Governance*, Oxford University Press: New Delhi, 2012, pp.42–43.

77 J. Srinivasan, No.70, p.34.

78 Potsdam Institute for Climate Impact Research and Climate Analytics, *Turn Down the Heat: Climate Extremes, Regional Impacts and the Case for Resilience*, A Report for the World Bank, June 2013, p.120. Available at http://www.worldbank.org/content/dam/Worldbank/document/Full_Report_Vol_2_Turn_Down_The_Heat_%_Extremes_Regional_Impacts_Case_for_Resilience_Print%20version_FINAL.pdf Accessed on July 24, 2013.

79 Avaya K. Nayak, "Post Super Cyclone Orissa: An Overview", *Orissa Review*, October 2009, p.98. Available at http://orissa.gov.in/e-magazine/Orissareview/2009/October/engpdf/Pages98-104.pdf Accessed on July 28, 2013.

80 Anurag Anamitra Danda, Gayathri Sriskanthan, Asish Ghosh, Jayanta Bandyopadhyay and Sugata Hazra, *Indian Sundarbans Delta: A Vision*, World Wide Fund for Nature-India: New Delhi, 2011, p.32.

81 Sujata Byravan, Sudhir Chella Rajan and Rajesh Rangarajan, No.101, p.43.

82 "A Poor Relief Effort", *The Hindu*, December 29, 2004.

83 R. Kumaraperumal, S. Natarajan, S. Chellamuthu, S.S. Ganesh and G. Anandakumar, "Impact of Tsunami 2004 in Coastal Villages of Nagapattinam District, India", *Science of Tsunami Hazards*, Vol.26, No.2, 2007, p.95. Available at http://tsunamisociety.org/262Kumarap.pdf Accessed on January 3, 2015.

84 Ministry of Environment and Forests, Government of India, *India's Second National Communication to the United Nations Framework Convention on Climate Change*, 2012, p.142. Available at http://envfor.nic.in/downloads/public-information/India%20Second%20National%20Communication%20to%20UNFCCC.pdf Accessed on July 26, 2013.

85 Atul Thakur, "Floods Affect 30 Million Indians Every Year", *The Times of India*, June 20, 2013.

86 Vandana Shiva et al., *Ecology and Politics of Survival: Conflicts Over Natural Resources in India*, SAGE: New Delhi, 1991, p.109.

87 "Assam Flood Toll Rises to 121", *The Hindu*, July 7, 2012.

88 Indian Network for Climate Change Assessment (INCCA), *Climate Change and India: A 4X4 Assessment, Sectoral and Regional Analysis for 2030*, November 2010, p.80. Available at http://www.moef.nic.in/downloads/public-information/fin-rpt-incca.pdf Accessed on August 1, 2013.

89 "How Snowmelt Fed Swollen Rivers", *The Times of India*, July 3, 2013.

90 Gautam Siddharth, "Restoring the Himalayas", *The Times of India*, July 4, 2013.

91 Potsdam Institute for Climate Impact Research and Climate Analytics, No.79, p.120.

92 Reshmi Kiran, Shrestha, Rhodante Ahlers, Marloes Bakker and Joyeeta Gupta, "Institutional Dysfunction and Challenges in Flood Control: A Case Study of the Kosi Flood 2008", *The Economic and Political Weekly*, January 9, 2010, Vol.XLV, No.2, pp.45–53.

93 "Fresh Flood Threat to Krishna, Guntur", *The Hindu*, October 6, 2009.

94 Ayesha Jain, "Displacement Explained: How Many Climate Refugees Does India Have?", *The Quint*, August 11, 2021. Available at https://www.thequint.com/climate-change/explainer-who-are-climate-refugees-and-why-migration-is-rampant#read-more Accessed on January 22, 2021.

95 Architesh Panda, "Climate Refugees: Implications for India", *Economic and Political Weekly*, Vol.XLV, No.20, May 15, 2010, p.77.

96 Aromar Revi, "Climate Change Risk: An Adaptation and Mitigation Agenda for Indian Cities", *Environment and Urbanization*, Vol.20, No.1, April 2008, p.210.

97 Mahendra P. Lama, "Internal Displacement in India: Causes, Protection and Dilemmas", *Forced Migration Review*, Vol.8, p.25. Available at http://www.fmreview.org/FMRpdfs/FMR08/fmr8.9.pdf Accessed on August 28, 2013.

98 Government of Bihar, *World Bank Global Facility for Disaster Reduction & Recovery, Bihar Kosi Flood: Needs Assessment Report*, 2010, pp.44–46. Available at http://www.gfdrr.org/sites/gfdrr.org/files/publication/GFDRR_India_PDNA_2010_EN.pdf Accessed on August 28, 2013.

99 "PM Calls as Assam Floods Worsen", *The Times of India*, July 9, 2013.

100 Sugata Hazra, Climate Change Policy Paper II, *Climate Change Adaptation in Coastal Region of West Bengal*, p.1. Available at http://awsassets.wwfindia.org/downloads/climate_change_adaptation_in_coastal_region_of_west_bengal.pdf Accessed on February 21, 2015.

101 Caesar Mandal, "Exodus from Tide Country to Kolkata", *The Times of India*, June 1, 2009.

102 "Laila: 6 Andhra Districts Hit, 50,000 Evacuated", *The Indian Express*, May 21, 2010.

103 McGranahan, G.D. Balk and B. Anderson, "The Rising Tide: Assessing the Risks of Climate Change and Human Settlements in Low Elevation Coastal Zones", *Environment and Urbanisation*, Vol.19, No.1, 2007, p.27. Available at https://sustainabledevelopment.un.org/getWSDoc.php?id=2393 Accessed on February 21, 2015.

104 Ministry of Environment and Forests, Government of India, National Communication (NATCOM), *Vulnerability Assessment and Adaptation, Chapter 3*, India's Initial National Communication to the UNFCCC, 2004, p.114. Available at unfccc.int/resource/docs/natc/indnc1.pdf Accessed on February 21, 2015.

105 Sugata Hazra and Anji Bakshi, "Environmental Refugees from Vanishing Islands", in Purusottam Bhattacharya and Sugata Hazra (eds.), *Environment and Human Security*, Lancer's Book: New Delhi, 2003, pp.220–221.

106 Ibid., p.223.

107 Architesh Panda, No.96, p.78.

108 Sarfaraz Alam, "Environmentally Induced Migration from Bangladesh to India", *Strategic Analysis*, Vol.27, No.3, July-September 2003, p.424.

109 Saleem Samad, "Refugees in Political Crisis in Chittagong Hill Tracts", in Omprakash Mishra (ed.), *Forced Migration in the South Asian Region: Displacement, Human Rights and Conflict Resolution*, Centre For Refugee Studies, Jadavpur University: Kolkata, 2004, pp.289–290.

110 Imtiaz Ahmed, "Teesta, Tipaimukh and River Linking: Danger to Bangladesh-India Relations", *Economic and Political Weekly*, Vol.XLVII, No.16, April 21, 2012, pp.52–53.

111 Ramaswamy R. Iyer, *Water: Perspectives, Issues, Concerns*, SAGE: New Delhi, 2003, pp.125–126.

112 Sanjeev Khagram, *Dams and Development: Transnational Struggles for Water and Power*, Oxford University Press: New Delhi, 2004, pp.45–49.

113 Annu Anand, "The Perils of Progress", *The Hindu*, March 22, 2013.

114 Ramananda Wangkheirakpam, "Lessons from Loktak", *The Ecologist Asia*, Vol.11, No.1, January-March 2003, p.19.

115 Dr. R.K. Ranjan Singh, "Tipaimukh Is a Death Trap for Indigenous People", *The Ecologist Asia*, Vol.11, No.1, January-March 2003, pp.76–77.

116 Varghese K. George and Chetan Chouhan, "The Law Catches Up with Vedanta", *The Hindustan Times*, August 25, 2010.

117 "Vedanta Project Fails Environment Test", *The Statesman*, August 25, 2010.

118 Minority Rights Group International, *State of World's Minorities and Indigenous People*, 2012, pp.135–136. Available at http://www.minorityrights.org/download.php@id=1112 Accessed on February 15, 2015.

119 Neil Padukone, "Climate Change in India: Forgotten Threats, Forgotten Opportunities", *Economic and Political Weekly*, May 29, 2010, Vol.XLV, No.22, pp.47–48.

3 Global Response to the Environmental Degradation and Climate Change

Environmental change is a significant challenge in our age. The global impacts and implications of climate change have given rise to the wide consensus among global community that measures should be taken in order to combat the crisis. The governments, policymakers, scientists as well as civil society groups of different countries are now more concerned about this change in climate and its disastrous effects. The IPCC also assesses the impacts and effects of climate change. However, now the issue of climate change has been moved from being only scientific matter to being the most contentious and distinct issue in global environmental politics.

1. The UN Response to Environmental Change

Human-induced global warming leading to environmental change is a worldwide problem today. The global response strategy to combat this problem is evident in various climate summits held under the aegis of the UN where meaningful attempts were made by the community of nations to curb GHG emissions. However, it is a challenge to reach unanimity as there exists contesting national interests to get a fair share in the atmosphere.

1.1. The International Environmental Change Regime

As far as international collective actions for saving our surroundings are concerned, 1972 is a landmark year when the United Nations Conference on the Human Environment in Stockholm, Sweden, ended with a declaration containing a set of common principles to inspire the global community in the 'preservation and enhancement of human environment'. It also established United Nations Environmental Programme as the 'environmental conscience of the UN system'.[1]

1.1.1. The UNFCCC Process

During the crucial 20 years following Stockholm conference, the world witnessed myriads of significant global collective actions[2] to address environmental degradation and climate change. They draw the background of the

DOI: 10.4324/9781003271192-4

United Nations Conference on Environment and Development (UNCED) in Rio De Janeiro in 1992 or the Earth Summit. As its fallout, the United Nations Framework Convention for Climate Change (UNFCCC) came into existence. It came into action in 1994. The convention asked the countries (here parties are classified into two main groups on the basis of commitments: **Annex I** Parties include the developed northern countries with industrial affluence; **Non-Annex I** Parties consisting of southern developing countries) to commit actions following the principle of 'Common But Differentiated Responsibilities' (CBDR) to reduce GHGs at internationally agreed levels. The developed countries are provided with the responsibilities to give resources to developing countries to help them with their efforts to limit GHG emissions.[3]

The other achievements of the Rio summit include the adoption of Agenda 21[4] and the Convention on Biological Diversity.[5] There is no doubt that the summit introduced a forum to address both environment and development; however, by embracing the idea of CBDR, it reflected the differences in the perspective of 'ecological space' between the North and the South. Another bone of contention was the issue of Global Environmental Facility (GEF) that was intended to be the financing mechanism for Agenda 21. The Rio summit and its fallouts gave impetus to a series of landmark climate conferences that ultimately gave birth to the Kyoto Protocol in 1997. Table 3.1 highlights the landmark Conferences of Parties (COPs) pertain to Kyoto and post Kyoto processes and post-2020 process.

Table 3.1 Landmark Conferences of Parties (COPs) Pertain to Kyoto and Post Kyoto Processes and Post-2020 Process

Year	Landmark Climate Conferences/Events
1997	Adoption of Kyoto Protocol at the CoP 3 which asks developed nations to reduce their emission of Greenhouse Gases (GHGs) to an average of 5.2 per cent below the 1990 level and to reach the goal between 2008 and 2009[6]
2005	Kyoto Protocol became operational
2006	Asia-Pacific Partnership on Clean Development and Climate (APP) had been launched involving six countries of the Asian Rim and Canada that joined later and the Major Economies Forum (MEF)
Post Kyoto Process	
2007	• Post Kyoto process started at the Bali Summit • Bali Action Plan had been adopted where the North–South divide (Bali Firewall) found its highest expression as it asked the developed countries that are not owing allegiance to Kyoto Protocol to make efforts for mitigation and also obligated the developing countries to adhere to feasible mitigation actions at the national level. Developed nations would provide financial and technological support to enable such actions which are also subject to measurement, reporting and to verification by the donors

(Continued)

Table 3.1 (Continued)

Year	Landmark Climate Conferences/Events
2009	• Copenhagen summit took place that came to an end with an agreement that was officially named Copenhagen Accord drafted by the United States and the BASIC (Brazil, South Africa, India and China). Both the industrialized nations and developing countries are asked to set their emission target by February 2010 • Regarding finance it had been decided in the accord signed in Copenhagen that developing countries would receive $ 100 billion as long-term funding. The poorest and vulnerable would be provided with short-term funding of $ 30 billion for mitigation and adaptation[7]
2010	Cancun summit that reached a compromise to set up a 'Green Fund' which was expected to mobilize $ 100 billion per year within 2020, for assisting developing states in adaptation and mitigation purposes
2011	'Durban Platform for Enhanced Action' had been launched at the Durban CoP, in the hope such a platform may develop a protocol with legal force applicable for 'all parties' that would blur the 'firewall'
2012	At Doha, the fate of Kyoto Protocol had been decided as parties agreed to its second commitment period (2013–2020) that would start right after the end of first commitment period, i.e. 31 December 2012
2013	A separate mechanism, namely, loss and damage, was proposed by the developing countries at the Warsaw CoP to address the climate liability of developed countries in addressing the damages already incurred by[8]
2014	At Lima CoP, an attempt was made for negotiating an outline text for the 2015 Paris agreement. The lima agreement includes a new phrase with CBDR & RC (Respective Capabilities) which underlined different national circumstances. The Bali 'firewall' was thus breached
2015	• A treaty on climate change had been adopted in Paris at 21st CoP which was a legally binding one • Paris agreement asks all nations to take an ambitious effort to limit the temperature increase in this century to 2 degrees and if possible even further to 1.5 degrees • It asks to review countries' carbon reduction pledges every 5 years based on each country's updated plan for Nationally Determined Contribution to the UN • Financial support to developing countries for mitigating and adapting to climate change was again underscored

Post 2020 Process

Year	Landmark Climate Conferences/Events
2016	Paris agreement came into force with 192 parties
2019	• The demand for linking climate finance to 'Loss and Damage' had been resurfaced at the Madrid CoP
2021	• Glasgow climate pact has been adopted at the 26th CoP which approves the rules of the Paris Agreement • Commitment of climate finance by the developed countries to mobilize a total of $100 billion per year starting from 2020 up to 2025 to help the most vulnerable • It has been agreed to double the proportion of climate finance going to adaptation • It identifies the need to stop the use of coal. In the pact after the final changes proposed by India, the phrase 'phase down' the use of fossil fuels instead of 'phase out' has been incorporated • It pointed out the need to reduce the inefficient use of fossil fuel subsidies

1.2. The Resultant Eco-Politics

The various climate summits efforts are made to reach meaningful actions to combat climate change, but they have ended with empty proclamations and blame game. The following fault lines exist in the climate regime:

1   The North–South divide on the question of assigning responsibility for cleaning up the polluted environment and over industrial countries' financial, technological or administrative support to the developing South;
2   The transatlantic strain between the EU and the US and disagreements within the developed block;
3   Division in the developing block on the question of responsibility and vulnerability.

Apart from these broad divisions, there exist other political alliances and grouping depending on the national interest of each party.

1.2.1. The North–South Divide

The centrality of the North–South divide is quite well known in various climate negotiations, which is surfaced the sensitive question – who will pay for the environmental mess? This dilemma lies at the heart of the environmental change problematic and is considered the bedrock of almost all climate negotiations and treaties aimed to establish equity in burden sharing. Although the South has caused little to the problem, it is not acceptable to allow them to pollute until reaching a certain level of development. Developing countries like China and India are among the top 5 emitters also. So in the post Kyoto process, developing countries were forced to take more obligations for mitigating climate change while looking for a more clean growth trajectory. However, it is iniquitous to blame them alone while the developed G20 countries are responsible for around 75 per cent of global emissions and in comparison to developing countries, their per capita emission level is also very high. The Paris Summit of 2015 brought the developing countries into the agreement to reduce 'emission intensity' of GDP of them, whereas the Glasgow went one step forward as it equalized both the developed and developing nations by asking them to reduce emissions to net zero.[9] However, these targets are voluntary and many are also conditional upon availability of adequate financial support which was another point of discontent.

The developing countries have put up resistance against a market-driven financial contribution which depends on the emission reduction obligations as taken up by the Annex 1 parties from the very beginning. To them, such financing must come as payments and should be in the nature of net transfer of fund that is grants not loans as well as should follow the priorities of the countries who are at the receiving end. The situation has further worsened as much of the promised financial assistance remains an empty shell. The promised $100 billion by 2020 has not yet been fulfilled. The rich nations collectively agreed to this goal but country-wise pledges were not made depending on their size. The United States failed to act in this regard though Japan and France paid more than their fair share but in the form of loan. During COVID-19, such climate finance also stalled which was a bad instance also.[10] In Glasgow summit of 2021, the demand for climate finance from the poor countries has been resurfaced. The UK pledged £290 million to help poor countries adapt to the impact of climate change at this summit.[11]

The lack of clarity regarding technology transfer and monopoly over eco-friendly technologies is also criticized. It helped the developed countries in maximizing their gain by manoeuvering the terms of transfer and also helped them in expanding the market for these technologies. The conflict between mitigation and adaptation has also complicated the relation between them. While the developing countries are in favour of adaptation than mitigating it through various technical and policy options, the developed North has given less priority to adaptation in fear that it would demand more liability on their part. Until and unless such problems can be properly mitigated, no climate deal modelled on a more inclusive version of Kyoto approach is feasible. So more adaptation funding is demanded from the developing countries. Many least developed countries also lamented that most of the finance are going to the middle-income countries rather than to the neediest ones. The V-20, a group of finance ministers from 48 climate-vulnerable countries, tabled another plan for $500 billion over 5 years as the $100 billion climate finance goal had not been met too.[12]

Despite this kind of split, during the Glasgow summit, decision to finance more for adaptation was a significant advance in the face of polarized positioning of major actors in climate imbroglio. The poor vulnerable countries of the South are also demanding compensation from the rich nations for loss and damage in various summits. In Glasgow, it has become a much debated issue but rich nations especially the United States and the EU are very reluctant in this in order to avoid additional financial liabilities. The G77+China pushed for a 'Loss and Damage Facility'[13] in the final agreement, but it was omitted which again sharpened North–South climate rift. However, the clash of interest is not only limited to North–South divide but also there have been changes in the strategic alignments of major groups in both. The reason is that countries are engaged more in promoting their respective national interests, though they belong to many different groupings and coalitions.

1.2.2. *Rift Within the Developed Block*

Since the negotiations that preceded the adoption of Kyoto Protocol, differences come into prominence between the EU and the United States.

While the European Union has politically been a strong supporter and fervent promoter of the protocol by pressing for deep cuts in emission levels, the United States is a stern detractor demanding more generous levels of emissions. The deadlock in the Hague conference was due to the clash between these two major actors where they were in disagreement on cost issue. The EU was reluctant to allow unrestricted trade and promoted a 'moral responsibility for national emission ceilings so that each country should reduce 50 per cent of its reduction commitment nationally'.[14] They did not want that a country should meet its target level by buying all the needed quotas from hot air holdings[15] of other countries. The desire was to cap the unlimited trading of credits. Moreover, the EU also opposed to accept the American claim for allowing the use of 'carbon sinks' in forests and agriculture as there is uncertainty in the exact amount of carbon removed.

The situation became more strained when the United States withdrew from the protocol. While such non-commitment to any binding treaty by the largest polluter helped the EU to adopt a leadership role in the debates and negotiations for climate change, it set them on a clear course of conflict as well. As the EU wanted a legally binding climate agreement with the United States on board, it vehemently protested such move of non-commitment to the existing climate regime. However, positive efforts were also made to change the situation, like the Gothenburg Summit where leaders from both the countries agreed to work in coalition to address the climate threat.[16] But they at the same time emphasized the significance of multilateral approach as the best option to combat climate threat, whereas the Bush administration was fundamentally opposed to multilateralism. Domestically, despite Bush's brusque rejection of the Kyoto, many companies were actively involved in reducing emissions. Even the Congress had adopted a resolution that forced the Presidency to re-engage in the international process. Thus, the United States faced both the internal and external pressures demanding substantial action from its side.

EU action was also stung by the US criticism that there was much talk but little action concerning Kyoto. In the post Kyoto negotiations, the problem has become more acute. There is failure to achieve consensus regarding the starting year against which emission reduction target by 2020 would be measured. While the United States along with Australia and Japan prefer a 2005 benchmark, 'many of the European countries like Germany and the United Kingdom insist on 1990 as the baseline'.[17] Not only that there were disagreements over setting targets as well. Before Copenhagen the EU had shown the most enthusiasm in reducing her emissions on the condition that other big polluters like the United States, Canada and Russia must declare ambitious commitments. But these countries took protectionist attitude in setting short-term target by 2020. The great reluctance for bearing the cost of implementing carbon emissions reductions by the developed countries became evident when Canada withdrew from the protocol in 2011. Japan and Russia have also opted to disregard their treaty obligations by rejecting to adopt target for the next period after 2020. Even the EU commitment to reduce 20 per cent of its emission from the 1990 baseline is insignificant since its emissions in 2009 were already 17.4 per cent lower than the base line.[18]

After Copenhagen, while the EU still had faith in an internationally binding treaty, the United States was harping on legally binding commitments based on domestic legislation. The United States under the Obama administration took more proactive role and emphasized strong domestic measures to cut emissions. The 'Clean Energy Act' was a milestone in this regard. Regarding the Southern demand for an agreement based on per capita emission line they also have different opinions – while the EU was not that antagonistic to this demand, the US opposed the idea vehemently fearing the loss of its pre-eminent economic status to India and China. Therefore, a reconfiguration of international climate regime and leadership crisis, with the EU unable to lead any more, had come into prominence after Copenhagen. Basically, the projection of power by the emerging economies since then, more specifically the BASIC which in coalition with the United States drafted the much criti-cized Copenhagen Accord and the US leadership aspirations, had actually started sidelining the European powers that had played the most significant role in the negotiations since the very beginning. The US diplomatic move to work in coalition with India and China in order to open an alternative track to Kyoto added fuel to it. Regarding climate finance, the EU scored a better position as it contributed more than the recent pledges that accounts for €23.39 billion ($27 billion) of climate finance in 2020. Along with this, President von der Leyen recently also proclaimed an additional € 4 billion as climate finance until 2027.[19] However, funding for green innovation from American side was not satisfactory. During the Obama administration, the US and EU cooperation on climate front was quite better than the later US administration. In such a situation, the EU has emphasized more in devel-oping climate cooperation and joint programmes with China which was evident in their 2018 commitment to cooperation on the implementation of the Paris Agreement.

1.2.3. The US–China 'Duopoly' and Climate Change

With the Copenhagen summit, the concept of G-2 evolved as a 'global axis' in the climate arena as without the engagement of the planet's two top emit-ters of GHGs, China and the United States, nothing decisive would come into play. However, each of them has used the other's non-compliance to Kyoto as an excuse for not acting. By signing the Bali Action Plan, although the United States started attempting to re-engage itself in the UN climate change process, simultaneously its efforts to open up alternative avenues like APP and MEF had shown its leadership aspirations. Both China and India became members of these groups. But they always have vehemently confronted the Western pressure for mandatory emissions cuts. Furthermore, since the APP countries shared a common vision to counter climate change without compromising the economic growth',[20] China had viewed the fora as a strategic means to ensure economic growth and social development while fighting climate change. At the same time, she does not want to lose her position as a 'de facto leader' acting in concert with the G-77 in the UN climate change negotiations and also as a member of the BRIC group to coordinate climate and energy policies. In recent times, China's effort to improve its energy efficiency at both the national and sub-national levels is remarkable. It also underscores the significance of renewables. India's National Action Plan was also an initiative in this direction. So their positions cannot be equated with that of the United States which is doing minimum at home while continuing to be lukewarm towards multilateralism.

The year 2014 had witnessed a major shift in the Sino-US relations with respect to climate diplomacy. These two biggest emitters have made an agreement, whereby the United States agreed to reduce 26–28 per cent of its emission in 2025 from its 2005 level, while China also agreed to lessen its emissions by 2030. Both these agreed pledges came immediately after the declaration made by the Europeans for 40 per cent emission reduction below 1990 levels by 2030.[21] During the Glasgow summit of 2021, again a joint declaration was made by both the countries announcing their willingness to work together on a number of climate-related actions to reduce emission, to tackling illegal deforestation and to increase their efforts for meeting goals of the 2015 Paris climate agreement.[22] The joint declaration gave assurance from both sides that together they would help in the transitioning to a net-zero global economy. Both the parties announced ambitious goals individually as well. However, there remains a significant gap between the ambitious pledges, their total effect and the actual required actions. Both the sides emphasized their willingness to act individually, jointly and with other parties, depending on the national circumstances, but more cooperation are due to close the gap. So, though the Sino-US climate cooperation in recent summits is a significant positive step towards mitigating climate change, it has many shortcomings and there exist a wide gulf between what they pledged and what is actually required.

1.2.4. *Fragmentation Within Developing Block*

Though they share common views regarding technology transfer and the additional financing for mitigation and adaptation and are siding with each other to strengthen the bargaining position vis-a-vis the developed world, there are deep differences within the developing block as well. The southern clans are in specific disarray subsuming several subgroups. The formation of the BASIC group just before the Copenhagen summit was an instance. Although they are part of larger G77 collective, together with South Korea and Saudi Arabia, they were responsible for about three-fourths of all emissions from the non-Annex 1 countries.[23] They were thus different from them. However, as they are not homogeneous sets of economies, there are internal divergences within their own ranks. These divergences are reflected in their climate diplomacy. They differ in their opinions about the definition of the term equity. While India has viewed equity in terms of per capita emission line combined with historical responsibility of the North for emissions, the other three countries with comparatively higher per capita emission have emphasized CBDR as the key to interpret the principle of equity. The Durban conference had projected the bifurcation within the group well. South Africa and Brazil were willing to accept binding commitments in return for finance and other concessions in principle.[24] China also indicated some 'flexibility' by offering to accept commitments on some conditions like amending Kyoto Protocol to cover all major northern emitters, operationalization of the promised $100 billion Green Climate Fund for developing countries soon and so on. India, however, held out against a new legally binding instrument and she raised her voice in favour of equitable distribution of global climate space. In doing so, she found herself isolated within the BASIC block and from other subgroups of G77 like AOSIS and LDCs. In such a situation, the European Union built a negotiating block with the AOSIS and the LDCs in order to isolate and divide the BASIC. The Union with the support from South Africa and many island states pushed forward the agenda of the Durban Platform. The summit witnessed the triumph of the EU climate diplomacy which had also received support from the United States as well, as these efforts would bring China and India under the purview of mandatory cut in emissions.[25]

The emerging economies in general and the BASIC in particular were blamed within their own natural constituency – the developing south. The Small Island States lamented that 'while they develop, we die' as they are facing existential threat due to climate change. They have underscored the difference between fight for survival and demand for equity in burden sharing. The latter in fact in coalition with the EU were in favour of imposing new legally binding emission limitation commitments from the emerging economies. The major oil-exporting countries who emit more than the LDCs and AOSIS countries, particularly the Arab members of the OPEC, have their own agenda as they consider themselves to be included in the list of most fragile countries. The 'response measure' to climate change may affect their economy as it requires diminishing dependence on fossil fuels like oil. So they have projected their concerns in the various climate talks. At the Glasgow summit of 2021 for the first time, the reference to the role of fossil fuel in climate crisis has been included. Fierce opposition from oil-producing and coal-producing countries is mounted against this move, and it attracts opposition from those heavily dependent on fossil fuels consumption too. However, Indian move to replace the term 'phase out' with the term 'phase down'; the use of coal which was supported by China as well has been criticized by the southern countries. The unilateral change of the text by India at the final hour created dismay, and it was blamed for non-transparency.

The declared positions of the states from both the developed and developing blocks on the involvement of the Security Council in climate change are also different as mentioned earlier. While most of the developed countries have historical responsibilities for climate change along with small island states, and some of the G77 states were supportive of the security council role, Russia, China and some Caribbean states opposed envisaging Council's role in the issue and viewed it another attempt to encroach on matters successfully dealt by the General Assembly.

The debate in the UNSC regarding the security dimension of environmental change and the Security Council's role in it therefore are contexts of discontent among members, and they lack the enormous political will that is required at this juncture to formalize the role of the UNSC in mediating disputes between blocks of nations in climate bargaining. While high-emitting nations like the United Kingdom, United States and Germany along with the most vulnerable ones like SIDS are in favour of Security Council playing a role, the emerging economies are skeptical as they feared it would endorse shared responsibility to curb emissions rather than differentiated responsibility as envisaged by the UNFCCC. Such rift and growing fragmentations of opinions among the members of the G77+China on various climate issues have actually affected the bargaining position of the developing South.

Various summits and conferences on environmental change have thus well revealed the fact that climate regime is fractured, and countries' vested national interests precede the concern for environment. In the UNFCCC/Kyoto process, countries are trying to build a climate regime, and its main aim is to reduce or minimize the emissions of GHGs. The UNFCCC and other multilateral bodies are also vocal about other chronic threats that global warming and its consequent climate change perpetuate like resource scarcity, sea level rise, sudden disruptions of life due to extreme weather events and so on. The problem of water stress, deforestation and fall in agricultural yield may be induced by several reasons, but environmental change basically exacerbates them, thereby affecting different aspects of human security. The following sections will deal with how the global community is responding to these threats.

(A part of Section 1.2 previously published in Das, Satabdi, "Negotiating an Intractable Climate Deal: The Kyoto Process and Beyond", Jadavpur Journal of International Relation, December 1, 2013. doi. org/10.1177/0973598414535061)

2. The Global Response to Environmental Change Induced Resource Scarcity

Drastic climate change requires adaptation and mitigation efforts from the concerned global community and the urge to signify it not only as a scientific phenomenon but as an important issue of policy concern. In today's world, the agenda for combating the threats associated with resource scarcity has thus made itself felt at the global fora. The following subsections shed light on the global responses to the scarcity of various life-sustaining resources.

2.1. Response to Lack of Food and Arable Land

The agricultural impacts of environmental change are felt in very different ways from country to country depending on varied soil quality, regional climate differences and above all socio-economic conditions. However, the 2007 IPCC report identifies in general two different strategies of actions for addressing climate change impact on agriculture. First, 'society can alter agricultural production processes to accommodate the altered climate' and second,

> society can act to reduce green house gas emissions in an effort to mitigate (or limit) the extent of future climate change, with farming playing a role in this effort. Climate change will affect agriculture negatively where societies do not find ways to adapt.[26]

The foundational period of global climate regime in the 1980s had witnessed several significant meetings and conferences during which scientific concern about global warming started evolving and then certain agendas were set with which actually the science of climate change entered the realm of policy issues. In the WMO-UNEP-ICSU convened Villach Conference of 1980 and 1985, the fact was recognized that atmospheric CO_2 concentration could have profound effects on global ecosystems, agriculture and water resources. Specifically, the conclusions reached at second Villach conference in 1985 were pertinent to agriculture. The conference asked the governments to support investigations of the sensitivity of the global agriculture resource base with respect to direct effects of increases in atmospheric CO_2 and other GHGs; effects of changes in climate and probable combination of these. The Toronto Conference of 1988 also recognized that changes in climate would diminish global food security. To enhance food security, the national governments were urged to reduce agricultural activities that would contribute GHGs to the atmosphere and to make efforts for regional and subregional cooperation. Various international organizations like World Bank, FAO and UNEP were also requested to join hands managing sustainable agricultural systems.[27]

The UNFCCC/Kyoto process has also recognized the mitigation potential in agriculture. Article 2 of the UNFCCC underscores the significance of wiping out all conditions that might threaten food production.[28] This process was carried forward by the Kyoto Protocol of 1997 that set targets for reducing agricultural methane and nitrous oxide emissions in developed countries. In accordance with the UNFCCC Article 4, keeping in mind the concept of CBDR, all parties are expected to reduce emissions for the benefits of agricultural sector.[29] The protocol also made agricultural soil carbon sequestration a voluntary mitigation action for Annex 1 countries under the Land Use, Land Use Change and Forestry (LULUCF) category. In 2006, the IPCC established guidelines for GHG accounting in Agriculture, Forestry and Other Land Use (AFOLU), replacing former LULUCF guidelines. The LULUCF activities are important in global climate change mitigation. It can generate a cost-effective way to reduce emission either by planting trees or by reducing deforestation.[30]

The Kyoto mechanisms like CDM gave little space to land-based mitigation activities because of concerns about the uncertainties in emission estimates and lack of permanence. The CDM methodologies for agriculture have limited applications as the methods often require time-consuming field measurement and access to analytical laboratories. Lessons from CDM experiences nonetheless are relevant to mitigation to agriculture in terms of general project accounting, as well as afforestation and reforestation projects that involved farmers.[31]

Since 2007, agriculture has been a component of negotiations on two tracks of the UNFCCC. The Kyoto Protocol Ad hoc Working Group is one track where agriculture has featured in debates about emission targets and the scope of CDM, while the second track that was created during CoP at Bali, namely, Ad hoc Working Group on Long-Term Cooperative Action (AWG-LCA), established reporting of nationally appropriate mitigation actions and classified agriculture as a sectoral approach to mitigation. The AWG-LCA also started negotiations on REDD that has created policy attention to agriculture as a driver of deforestation. Additionally, it catalysed a technical work programme, a period of intensive review and development of technical options and finance for REDD that presents a model for what could be done for agriculture.[32]

The nexus between climate change mitigation and adaptation and food security was once again underscored in the World Food summit declaration on food security held in November 2009. In the Copenhagen Summit of 2009 and Cancun summit of 2010, informal discussions took place on the nexus between agriculture and activities pertaining to mitigation and adaptation. In Copenhagen, it became clear that most developing parties consider activities in the agriculture sector as important aspect of their NAMAs. The Accord also opened avenues that would define the scope of NAMAs while supporting mitigation actions that reduce emissions from agriculture and other land use.[33] To address the impacts of climate change on agriculture, a text had been drafted but it was not agreed upon. Durban Conference was significant in this regard as for the first time, the agenda for the Subsidiary Body for Scientific and Technological Advice (SBSTA) mentioned agriculture in its draft.[34] In the SBSTA in its different sessions like in Bonn 2012, different parties discussed future course of actions for mitigation and adaptation processes in the agricultural sector, issues pertaining to food security, etc.[35] An agreement was reached to accelerate adaptation efforts in climate change. It included rural sustainable development in all countries also.[36] Different parties to UNFCCC responded differently regarding the agriculture-related work programme. The USA proposed increased funding in the said sector in any post-2012 agreement and suggested some broad issues to be negotiated within SBSTA. These issues include balancing mitigation with adaptation activities, steps to augment production, resilience building and transferring technology. Japan emphasized the role of SBSTA to act on devising mechanism for carbon sequestration in agriculture, technology transfer and knowledge exchange, etc. In order to include carbon sequestration project in agriculture, France proposed to extend CDM.[37]

The mitigation–adaptation dilemma is a very common issue while solving the problems associated with climate change and food production. Developed and developing countries are of different opinions regarding this in Doha. Unlike the developed parties, developing ones demanded to keep agriculture within the perimeter of adaptation only. The G-77 collective along with China envisaged that as part of mitigation effort, if agricultural emissions are counted, then it would enter the realm of carbon market and exploitation would follow in the agricultural sector too.[38] At Doha, developing countries also voiced concern for the smallholders and marginal peasants and urged for safeguarding their interests as part of adaptation measures.[39] So it can be said that the developing countries are skeptical about the Northern move to include agriculture as part of mitigation actions and have prioritized adaptation measures to combat ill-effects of climate change in agriculture.

Given the environmental change triggered food scarcity, there is growing consensus among the developing countries that agriculture sector should not be considered for emissions reduction only, but the need for the age is to make the crop production system more sustainable and climate resilient. Sometimes, use of biotechnology in agriculture and the GM crops is considered a solution to scarcity of food. But it might strengthen the North–South divide again. The farmers are forced to depend on big agribusiness companies, monopolized by developed countries and to open avenues for exploitation. In such a situation, the impoverished South requires proper assistance from developed world in the form of technology transfer, economic resources required for capacity building for addressing agriculture and climate change adaptation and mitigation. Although both the industrial model of agriculture practised by developed nations and conventional farming that is prevalent in the developing world are contributing to climate change problems, the smallholder farmers in developing countries, notably, the LDCs, are the most vulnerable to its adverse effect. Against such a backdrop what they require the most is the sustained and large-scale food security adaptation initiatives.[40]

Agroforestry which is a LULUCF activity is another option. It can fulfil various on-farm adaptation needs which helps in increasing agricultural productivity.[41] This system is gaining popularity because of the fact that smallholders can reap co-benefits from it. It may generate social, economic and environmental benefits and services if the right practices are used and could be promoted in development programmes to benefit poor rural households, mainly subsistence farmers on small landholding.[42]

Food scarcity has also been augmented by land degradation and desertification. In the Rio Earth Summit, climatic variations were considered one of the major reasons behind desertification. The United Nations Convention to Combat Desertification (UNCCD) 1994 also adopted the Earth summit's definition of desertification. Given the intensity of this problem, combating desertification becomes essential in order to ensure the long-term productivity of inhabited drylands. Table 3.2 shows the various steps that are taken by the global community to combat desertification.

Table 3.2 Steps to Combat Desertification

Year	Measures taken at the UN
1974	The UNGA by resolution 3202 suggested that the global community to act responsibly to arrest desertification and make plans for the economic and developmental recovery of the affected regions
1977	The United Nations Conference on Desertification took place in Nairobi. It ended with a Plan of Action consisting of 28 measures that national, regional and international institutions could take to halt land degradation around the world
1984–1991	The UNEP had taken many steps to evaluate the implementation of the Plan of Action in countries with land at risk which was designed in the United Nations Conference on Desertification (UNCOD)
1991	Adoption of the Abidjan Declaration by 40 African states who assembled for a regional preparatory meeting for UNCED. It called for a convention to combat desertification
1992	At the Rio Earth Summit, the UN called on to negotiate an international legal agreement on desertification
1994	The United Nations Convention on combating desertification came into force in 1996. It provides a detailed and comprehensive roadmap for combating desertification. All the Gulf Cooperation Council countries signed and ratified the convention
2006	The United Nations launched International Year of Deserts and Desertification
2007	The adoption of 10-Year Strategy of the UNCCD (2008–2018) that endorsed global partnership to stop desertification
2010	The UN Decade for Deserts and the Fight Against Desertification (2010–2020) had been launched by the UNG that had the goal to protect the dry lands
2015	Sustainable Development Goals 15 of 2030 agenda underscored the importance to strengthen cooperation on desertification

Source: Data collated from *Kannan, Global Environmental* Governance and Desertification: A study of Gulf Cooperation Council Countries, Concept Publishing Company: New Delhi, 2012; "Desertification and the UN System", Earth Negotiations Bulletin, available at http://www.iisd.ca/vol04/0401018e.html Accessed on 18 November 2014; United Nations, Desertification, Land Degradation and Drought, May 14, 2018. Available at Desertification, land degradation and drought. Sustainable Development Knowledge Platform (un.org) Accessed on January 19, 2022.

Various states that witnessed desertification now are working towards the implementation of the UNCCD for carrying out several action programmes. Countries have now become more conscious of this problem as if it goes unnoticed the end result would be the complete transformation of fertile land into wasteland incapable of yielding anything. Additionally, food security and agricultural productions are inextricably linked with water resources. As unsustainable and overexploitation of water and climate change have created a serious imbalance in the supply and demand of water worldwide, adequate measures are required to confront water scarcity related problems. Otherwise, the crisis triggered by food insecurity will be more aggravated, and solutions to the problem may remain more intractable.

2.2. Response to Water Scarcity

One of the most crucial ways to adapt to the climate change induced negative consequences is to manage water effectively as it is badly impacted by changes in the environment. The management of this resource should therefore require great concerns from all. There are two crucial aspects of the problem that the global community should consider while addressing environmental change induced water stress: First, how to administer the sustainable use of fresh water resources; Second, the debatable issue of transnational water bodies which straddle national borders. In both cases, proper water governance is required that demands actions from different actors at all levels – global, national and local. At the global level, there were several programmes undertaken by the UN which has considered water issue as one of the vital components of discussion in various climate-related summits and conferences. Such a conference took place in Dublin, Ireland, on 26–31 January 1992, namely, International Conference on Water and the Environment (ICWE). The participants gathered here emphasized finding fundamental new approaches for assessing, developing and managing fresh water resources.

In the Earth summit, the approach of integrated water resource development and management (IWRM) was agreed upon by the parties. The final resolution from Rio, 'Agenda 21' deals elaborately with water resources, its scarcity, gradual destruction and aggravated pollution. It also noted the impacts of global warming on the nature and availability of fresh water resources.[43] The Agenda 21 of Rio Earth Summit also emphasized the management of water in an integrated way bringing different users and sources of water together.[44] Discussions and actions taken to deal with the issue of water scarcity in Dublin that was consolidated into Agenda 21 were reaffirmed in 2002 by the Johannesburg World Summit on Sustainable Development (WSSD). It had also been outlined that short-term target for Integrated water management as well as water efficiency plans must be adopted by 2005.[45] For the follow-up process of the WSSD, UN-Water was endorsed in 2003 which is an inter-agency entity of the UN. In the Rio+20 (UNCSD) that was held 20 years after the 1992 Earth Summit, water remained one of the priority areas in this conference.[46]

There were other efforts to govern water-related actions on a global scale. The formation of Global Water Partnership (GWP) in 1996 was a response to this initiative. This tried to bring all organizations that were engaged in water resource management and underlines the significance of ensuring cooperative and coordinated water management.[47] There are other manifestations of this sort like World Commission on Dams, International Rivers Network, World Water Council and so on. The first World Water Forum (WWF) was held in 1997 at Marrakesh. It was formed triennially by the World Water Council.[48] In March 2000, delegates from different countries who have expertise in managing water assembled in the Hague for the second World Water Forum. This was organized around two important reports ('The World Water Vision' and 'World Water Security: A Framework for Action') in order to present an authoritative frame for addressing global water problems and its solutions. These two reports talked about the privatization of water.

Rivers are important sources of water. But often river catchment areas are undergoing severe change due to human encroachment resulting in climate catastrophes. The problems are more acute when cross-border river bodies are concerned about whose shared water is creating disputes among co-riparians. Although there are many treaties and agreements to resolve these transnational water disputes, they are inadequate with respect to responding to the ecological or political challenges posed by conflict over shared fresh water resources. But climate change is rarely discussed in these transboundary water agreements. They are mostly unaware of the fact that future water supply and quality may be altered due to changing climate, hampering the reasonable sharing of water among the riparian states. Not only that managing shared water resources in a sustainable way during climate change is also challenging as multiple political entities and actors involved may differ in their respective views. Often their attitude may be guided by the existing political fractions as well. The governments of co-riparians therefore before signing any treaty regarding water sharing should take into account the fact that upstream dam building and river diversion may result in water scarcity downstream. Such changes in the hydrology of the transnational river basins by hydro-engineering are often guided by geopolitical interests of respective countries too. But they forget that environmental hazards do not respect borders and any kind of human interference to the natural system is catastrophic affecting the interests of all the riparians of shared river basins.

Given its multidimensional significance to human well-being, climate-induced water scarcity is thus considered vital for global response. Effective environmental governance regarding water scarcity requires efforts to devising methods to address deforestation too as they are inextricably linked. The following section will throw some lights on how the global community responds to this problem as forests that act as either sinks or sources of carbon emissions are significant for both economy and existence of some communities.

2.3. Response to Deforestation

Conservation and sustainable management of forests have become a politically significant issue today. Tropical deforestation is clearly identified by the UNFCCC in its text. But the UNFCCC does not provide any mandate or incentive for reducing emissions from tropical deforestation. So in the initial years of UNFCCC process minimum steps were taken in this direction. The Rio Earth Summit ended with agreeing only to the Forest Principles that are non-binding in nature.

In international law, forests are a sovereign resource of the state. Some tropical countries asserted this view. Therefore, while developed countries of North America, Europe and Japan argued for a forest convention and countries like the EU, the United States, Canada and Japan voiced for linking the concept of sovereignty to two other principles of stewardship and common responsibility, all the developing countries of Asia, Latin America and Africa argued against. Malaysian delegation, speaking on behalf of the G-77 collective, retorted by stating that the concepts such as global commons had a supranational character and were an attempt by the North to erode the sovereignty of developing countries over their forests. Basically, the tropical forest countries to whom the rainforests are important for producing timber and carbon sinks tried to argue from the economic gain perspective and they were demanding that any agreement on forest conservation should be tied to debt relief, greater financial assistance and increased transfers of environmentally sound technologies. The Northern countries were not in favour of appreciating such demands and were likely to agree to significant North to South resource, and technology transfers only if they could draw some binding commitments from the South in forest conservation targets.[49]

As a result, the UNCED negotiations on forest were stalled by divisive political and economic interests of the North and the South. A confidence building dialogue had been initiated to break the impasse between Canada and the Malaysia. The fallout of that effort was the formation of Inter-governmental Panel on Forest (IPF) to provide a forum for forest policy deliberations. It met four times between 1995 and 1997 and negotiated a series of non-legally binding proposals for action. Subsequently, in 1997, the Intergovernmental Forum on Forests (IFF) was established by the ECOSOC for the period 1997–2000. For sustainable forest management, almost 270 proposals for action were made as part of the IPF/IFF process.[50]

Another similar attempt was made in 2000 to advance 'sustainable forest management'. UN Forum on Forests (UNFF) that involves a wide array of multilateral and bilateral agreements and initiatives regarding forest conservation came into being in that year. However, the demand for a common forest convention and opposition to this was common in various UNFF sessions as well. In 2005, UNFF tried to make countries agree to a new international forest instrument. But intense debate took place as countries differ in their opinions on whether it should be a convention or non-legally binding instrument. The Amazonian Pact countries like Brazil, Bolivia, Peru and Ecuador along with the United States that was not in favour of any international environmental commitments, opposed to convention. Brazil and the United States formed veto coalitions. As a result, states agreed to negotiate 'Non-Legally Binding Instrument on All types of Forest' which concluded in 2007.

In various climate summits, efforts were made to value and price natural resources through market mechanisms too. Here comes the issue of reduced emissions from deforestation (RED). It implies that countries which reduce their deforestation above a certain baseline will create carbon credits that they can sell to countries which wish to exceed their agreed emissions level in a post Kyoto market based global carbon trading scheme. If we start with the Kyoto Protocol of 1997 which is landmark in the history of climate negotiation, such scenario becomes evident. Article 12 of the Protocol, which introduces the CDM, itself neither promotes nor prohibits projects that reduce emissions from tropical deforestation. It allows only plantation projects – reforestation (planting forest in areas that were deforested before 1990) and afforestation (planting forest in areas where there was previously no forest vegetation for at least 50 years) – aimed at sequestering carbon from the atmosphere.[51] While Article 3.2 of the Protocol addresses emissions from deforestation, on methodological grounds, and because of sovereignty concerns raised in particular by Brazil, the Marrakesh Accords of 2001 excluded deforestation emissions from flexibility mechanisms. The LULUCF limits projects to afforestation and reforestation under the CDM, setting aside emissions from deforestation.[52] Despite international debate on this issue, CDM excludes the forest conservation projects from its ambit. Therefore, countries like Brazil and Indonesia were somewhat disincentivized to participate in the Kyoto efforts as they found no specific provisions or financial incentives to do that.

The year 2005 was a turning point as during the Conference of Parties' 11th session held in Montreal, Papua New Guinea and Costa Rica proposed to put on the agenda the issue that talked about compensation for developing countries for reduced deforestation. Coalition for Rainforest Nations (CfRN) supported the move.[53] With this, the effort to Reduce Emissions from Deforestation (RED) started entering the realm of climate negotiations. Papua New Guinea, Costa Rica and a handful of other developing states, now united as the CfRN thus reintroduced the idea of RED. Between 2006 and 2008, in the discussions under the UNFCCC process, the drivers of deforestation had been focused on. The other issues that had also been discussed included measures for monitoring emissions and the impediments both technical and financial in the way of its implementation.[54] However, the debate pertaining deforestation had been intensified in Nairobi in 2006.

The tropical forest countries were presenting conflicting views regarding incentives for reduced emissions.[55]

At Bali, the decision had been adopted by parties to stimulate actions towards reducing emission from degradation of forest and deforestation in developing countries. The term 'forest degradation' was added to RED here. In fact at CoP13 in Bali, RED was a standout issue and RED became REDD+ that extends the scope of RED to include the durable management and conservation of forest and carbon stocks. The scope of the discussion broadened eventually. It was a victory for some coalitions, such as the Commission des Forêts d'Afrique Central (COMIFAC, Central African Forests Commission), as well as for some countries (e.g. India) with low current deforestation rates (and thus with little room to earn credits from reducing emissions from deforestation), yet still having sizeable forest areas. The Bali outcome was a carefully crafted compromise between COMIFAC, India and others with an interest in conservation, and Brazil and others, which supported the original formulation.[56] During the period 2008–2009, incentives pertain to REDD in developing countries as well as approaches to conserve and manage forests were considered as part of the process of Bali Action Plan.[57] At the Copenhagen Summit concern for deforestation was once again expressed. REDD+ was the only mitigation option explicitly mentioned in the Copenhagen Accord.[58]

The Cancun Summit of 2010 started with the hope that it would finalize the agreement on REDD. But the outcome did not satisfy the expectations. The text of the agreement only emphasized the role of developed countries in providing financial support to developing ones. The developing states were asked to create national REDD+ strategies, including Reference Emission Levels[59] and transparent monitoring and information systems to demonstrate how environmental and social safeguards would be respected.[60] At the CoP17 at Durban, a decision was taken that focused on the financing meant for the result-oriented complete materialization of REDD+ activities. The term 'safeguard' had been discussed with emphasis in Durban in 2011. The safeguard principle implies the environmental and social integration process during the course of a REDD project. The forest countries were required by the final decision to report their process of implementation of safeguards.[61] REDD+ decisions were considered by many as disappointing because they did not clearly define the positive incentives associated with it.[62] Finally, at Warsaw talks 2013, the participating nations agreed for REDD+ mechanism. A work programme called Warsaw Framework for REDD+ was adopted here. Projects showing that they have taken measures to avoid deforestation would be financed under this programme. It had been further outlined that payments would be result based.[63] Article 5 of the Paris Agreement also recognized REDD+. Till January 2020, 50 developing countries submitted a REDD+ forest reference level for technical assessment to the UNFCCC.[64] At the Glasgow summit, the Coalition for Rainforest Nations launched the first REDD+ rainforest carbon credit auction in compliance with the Paris Agreement.[65]

However, REDD is not beyond criticism. Through this, the trade in forest carbon credits precedes the real need to protect the environment and may create perverse incentives to capture carbon funds. The developing countries may bargain 'generous deforestation baselines' before entering the agreement for this as they could claim higher level of reduced deforestation than the actual one. Developed countries can also get advantage through this as they are able to continue polluting the environment without stop emitting domestically. With market mechanism, they can buy other's carbon space or sequestration of it. Second, REDD only talks about the carbon stock value of forest. Apart from it, as a public good provider forest is significant for keeping biodiversity intact and it has value for producing timber. It did not consider the fact that deforestation may also result from illegal felling of trees. Moreover, it had been pointed out by the Indigenous people that their concerns are not taken care of in the international negotiations though they are the direct victims of deforestation.[66]

In the Glasgow summit, over 100 world leaders have pledged to end and alter deforestation by 2030. The governments of 28 countries also promised to remove deforestation that originates from the global trade of food and other agricultural products. Some biggest financial companies like Aviva, Schroders and Axa also committed to end investment in activities linked to deforestation. Finally for the protection of the tropical rainforest in the Congo Basin, decisions were taken to establish a £1.1 bn fund.[67]

The Indigenous population are the most vulnerable section as forests provide their food and shelter. Vested economic interests and commercial using of forests have a direct bearing on them. So, various declarations related to this community, such as the 2002 Kimberley Declaration, have emphasized the danger of climate change repeatedly.[68] In the Amazon, Climate Alliance had been established between the Coordinating Body of Indigenous Organizations of the Amazon Basin (COICA) and the peoples of European cities in 1990. It also talked about the rights of the Indigenous people and generated support for their effort to protect the carbon reservoirs in the tropical forests.[69] The Quito Declaration of 2000 echoed the same and endorsed the need to increase awareness of the Indigenous people. However, in spite of their growing awareness of climate change, the plight of the Indigenous community is rarely concerned in the popular discourse on climate change. In various climate summits, the Indigenous people's organizations like Indigenous Peoples Bio-cultural Climate Change Assessment Initiative (IPCCA) and Tebtebba Foundation are critical of the forest regulations, and they are of the opinion that concerns of Indigenous people are not sufficiently taken care of in the process of mitigation and adaptation. In 2011, IPCCA members opposed REDD+ as its implementation would act to the detriments of forest communities.[70] In 2021, COP 26 at Glasgow ensured finance to support Indigenous communities.[71] Thus, the global response to combat deforestation and other associated threats that affect the humanity severely has many shortfalls. But the positive sign is that the global community is recognizing the significance of forest as an essential carbon sink, as a repository of terrestrial biodiversity and in terms of its economic values. Therefore, the responses towards deforestation try to address those issues as well that have affected the forest-related goods and services.

2.4. *Response to the Problems Pertaining to Energy Resources*

In various climate summits, policy recommendations have been proposed and debated as a growing economy requires both affordable and sustainable supply of energy. However, such efforts have failed to achieve cooperation internationally on carbon emission reductions, and there is little agreement on the best suitable actions needed to reduce global dependency on fossil fuels.[72] Even combustion of fossil fuels and peak oil are hardly debated and discussed in the climate change conferences. Against such a backdrop, there exist growing interests in deploying energy sources and technologies that produce lesser amount of GHG emissions. Modern biomass sources for transport, heat and electricity, hydropower, solar, wind and other renewable energy sources have increased their contribution to the global energy balance at remarkable rates.

There exist various technical options to reduce carbon dioxide emissions from energy sectors. Carbon Capture and Storage (CCS) is one of them which is a key tool in tackling climate change by decarbonizing the energy system. By using CCS, carbon dioxide (CO_2) is captured during the burning of coal and gas for power generation and during the manufacture of industrial facilities.[73] If these systems can be developed as hoped, the emissions from a coal-fired power station could be reduced by 80–90 per cent, the gases produced being placed in aquifers or other geological formations.[74] But these options are not sufficient and face a number of hurdles. It is dependent on the price of carbon and the technological advancements.[75] The use of cleaner energy as an issue of environmental sustainability came into being in the late 1980s and 1990s with the spread of the modern environmental movement. Along with growing concern over global environmental change, from that time it also acquired growing momentum. However, there was no explicit section on energy in the Rio Declaration and the Agenda 21, but the linkages between energy and the environment were established.[76] The UNFCCC and its subsequent processes have also acknowledged that energy accounts for a large share of greenhouse gas emissions.

Apart from global declarations and conventions, the World Bank made effort to internalize environmental sustainability objective by articulating an environment strategy for energy in the 2000 Fuel for Thought document. In 1992, the Asia Alternative Energy Program was established by the Bank and its donor partners in order to assist Asia in bringing renewables and energy efficiency into the mainstream. It also launched a clean coal technology and the Solar Development Corporation to accelerate the use of solar photo-voltaic. All of these proved the Bank's activities in the energy–environment area.[77]

The Kyoto Protocol also identified the significance of climate–energy nexus, and it envisaged CDM for the promotion of renewable energy and energy efficiency as required by the mitigation efforts. Emission trading, another Kyoto flexible mechanism, also helps in establishing energy efficiency. European Union's Emissions Trading system (ETS) is the largest scheme of emission trading within the Kyoto regime. It was considered the first mandatory cap and trade system covering GHG emissions from industry. It had been used by the EU as a means to meet Kyoto Protocol's obligation.[78] Apart from landmark climate summits, energy-related environmental concerns have increasingly found their way into high-level political statements as well. The World Summit Outcome 2005 adopted by the General Assembly has vividly mentioned the energy issue and its relation with climate change. The UN-Energy was established after 2002 World Conference on Sustainable Development in Johannesburg as an attempt to bring together the assortment of UN bodies that play some role in energy, although it lacked substantial resources.

Along with the UN process, there were various other institutions that have broadened knowledge about the nexus between climate change and energy security and demonstrated extensive interest in mitigating the challenge. This was the reason behind G-8 heads of the state's plea at the G8 Gleneagles Communiqué of 2005 to the International Energy Agency (IEA) for devising possible ways for bringing a clean energy trajectory. The outcome document of the meeting, namely, Gleneagles Plan of Action (GPoA), on climate change, clean energy and sustainable development outlined measures in key areas such as energy efficiency, renewable energy or financing of clean energy projects.[79] The meeting also resolved to engage international institutions in the implementation of GPoA including the IEA for advice and for financing projects in developing countries, World Bank and other multilateral banks. Pledges were also made here from various developing countries for combatting the rising GHG emission due to fossil fuel combustion.[80]

The G8 leaders at St. Petersburg in July 2006 also were in favour of devising alternative energy scenarios and strategies. They formulated the 'St Petersburg Plan of Action on Global Energy Security' envisaging the importance of energy security.[81] The Pittsburgh G-20 summit in 2009 focused on promoting energy efficiency in particular by phasing out subsidies for fossil fuel, in part because they impeded a transition to clean energy sources, a concern echoed by the Leaders Declaration from the Asia-Pacific Economic Cooperation (APEC).[82] In 2007, the Major Economic Forum convened the first meeting on 'Energy Security and Climate Change' which focused on key areas like energy efficiency and low-carbon-intensive power generation. All the participants prioritized required actions for addressing energy security in their national statements.[83]

The 2008 G8 summit of Toyako was another milestone where again leaders reaffirmed the need to build energy efficiency as it was considered indispensable for ensuring energy security for acquiring economic and environmental goals.[84] The World Bank also drafted 'Clean Energy Development Investment Framework' to help countries, especially in Africa to decode the needed investment required for accessing energy, transitioning to low-carbon trajectory and adapting to the effects of environmental change.[85] Thus, in maximum global political declarations made by G8 or G 20 or MEF or World Bank, the explicit mention of climate change in particular becomes usual and they have underscored repeatedly the close interaction between the three parameters of energy, economy and ecology.

Along with such organizational efforts, the major economies also started implementing their own climate change programmes that accommodates their domestic policies regarding economic growth and energy needs while taking care of the environmental concerns. In 2007, the EU drafted a proposal which was adopted in December 2008, to cut GHG emissions by 20 per cent along with improving energy efficiency to 20 per cent and to ensure the availability of 20 per cent of energy from renewables by the year 2020. Later in 2014, the 2030 framework was adopted, comprising the improvement of energy efficiency by 32.5 per cent, 32 per cent increase in the use of renewable energy and cuts in GHG emissions. The Obama government in the United States also listed energy and climate change among priorities of his presidency with plans to reduce GHG emissions up to 17 per cent by 2020 and up to 83 per cent by 2050 against their 2050 levels. The Trump administration failed to continue this trend of climate consciousness, but later the Biden administration is scoring high in proclaiming ambitious climate actions. Even his campaign included the demand for enacting legislation to establish 'an enforcement mechanism to achieve the 2050 goal, including a target no later than the end of his first term in 2025.[86] China also adopted a National Climate Change Programme in 2007, determining targets for reducing energy intensity by 20 per cent below the 2005 level by 2010, while increasing 20 per cent of the share of renewable energy in electricity generation within 2020, and pursuing supportive policies.[87] The International Energy Agency in an report noticed that to meet the long-term goals of the Paris agreement, 2015, 15–65 per cent increase in the share of renewables by 2050 is required. The transition is ongoing in the power sector. But there exist bottlenecks in the capacity of electrical transmission in large markets such as Germany and China. It inhibits the expansion of renewables.[88]

As fossil fuel combustion contributes to global warming, the world has recognized the need to create conditions so that all can reap the benefit of clean and efficient energy. This was the reason behind the launching of Sustainable Energy for All Initiative by UN Secretary General Ban-ki-moon on 7 November 2011. The UNGA also declared the year 2012 the International Year of Sustainable Energy for All. Thus, energy has taken the central place in the multilateral processes.[89] The first annual Sustainable Energy for All Forum has opened on 4 June 2014, which sets the stage for the launch of the UN Decade (2014–2024). More than 80 developing countries have joined the initiative.[90] In this way, it has been acknowledged by the global community that energy efficiency and sustainable use of it can curb GHG emissions. Energy efficiency and renewable energy are vital to solve environmental problem spurred by the expanding energy needs of developed and developing nations. It can be deemed as a global response to the problem associated with energy–economy–environment complex.

The use of renewable energy sources is another possible way to mitigate the environmental security risks caused by energy as mentioned earlier in this section. The growing significance of these resources at the global level was reflected in the formation of International Renewable Energy Agency (IRENA) in 2009.[91] The IEA has projected that around 20 per cent of future emissions reductions could be achieved through the greater use of energy from renewable sources, in all its forms: electricity, heat and cooling, and transport.[92] Fuel-cell and nuclear energy are also significant in revising the energy system. Biofuels have the potential to replace significant quantities of gasoline, diesel and jet fuel, hence reducing dependence on oil too.[93] But despite such benefits, renewables are not widely considered a panacea for all the security risks posed by energy-driven climate change.[94] Sometimes, the concerns for relatively high cost along with physical constraints of connecting renewables to the grid and lack of pipeline capacity have inhibited the development of renewable energy. Many developing and emerging nations still face the dilemma of deciding between expensive investments in cleaner energy and using easily available fossil fuel. Fossil fuel subsidies that dwarf support for renewables have exacerbated the problem. Therefore, steps to phase out these subsidies that drive up carbon emissions are significant.[95] In the Glasgow summit of 2021, efforts were made to phase down the use of coal and reduce the fossil fuel subsidy.

The efforts to eliminate dirty energy resources in order to save the planet can become successful only if there exists effective international coordination. The fossil fuel based developing countries who contributed little to global climate change may find it difficult both economically and politically to reorient their coal-based energy policies in order to confront climate change. So the energy governance structure should incorporate emerging economies and major oil-producing countries in the decision-making procedures so that the national interests of all are maintained. For example, the fossil fuel based economies or the oil-producing countries may have different interests which should be taken into account and along with green technology, investment in carbon sinks are more urgent for them to sustain their developmental path. So without a comprehensive approach addressing energy needs, supply and sustainable use and production of it, no climate protocol or convention will be successful.

3. Global Response to Extreme Climate Conditions

3.1. *Response to Sea Level Rise*

There is no doubt that the risks from sea level rise are imminent and serious, and it would continue beyond 2100.[96] As the 52 SIDS are the most vulnerable to this threat of rising sea levels, coastal inundation and storm surges, most of them have formed a group called Association of Small Island States (AOSIS) in the 1990s that tries to uphold the interests of these countries in the global climate agenda. The Pacific SIDS (PSIDS) has also formed a caucus organization for representing its interests at the UN as well.

In Rio, in 1992, the international fora first identified the unique and particular vulnerabilities of SIDS and the latter was also instrumental in the drafting of Kyoto Protocol. Thirty-eight SIDS have ratified the UNFCCC and Kyoto Protocol. National Adaptation Programmes of Actions (NAPAs) were also submitted by 11 SIDS who are of the same status as LDCs. The Rio Summit recognized that due to the geographic dispersion, isolation, small size and narrow resource base of these countries, they require special mention in any environmental deal. Therefore, the special case of SIDS was incorporated in Chapter 17 of Agenda 21. Such consciousness also resulted in a UN resolution to have an International SIDS Conference in 1994. The AOSIS acted as a catalyst and took the initiative in organizing the global conference on sustainable development of SIDS which adopted the Barbados Programme of Action (BPOA) in 1994. Its aim was to ensure sustainable development of SIDS.[97] The BPOA identified 14 thematic areas for specific actions. Among them climate change and sea level rise are the most important as the plan considered it as one of the main challenges faced by the small island developing states.[98] BPOA had been updated a decade later in the Mauritius Strategy.

Before the Mauritius conference, another important interregional meeting was held in Nassau, Bahamas. The Nassau Declaration that was adopted reaffirmed the significance of BPOA for providing the main framework for sustainable development in the SIDS. It also ensured the commitments from the small islands to follow the targets and timetable of the MDGs, on climate change.[99] Against this backdrop, in January 2005, the International Meeting to Review Implementation of the Programme of Action for the Sustainable Development of SIDS was held in Port Loto be uis, Mauritius. It presented a significant opportunity to consider the best possible efforts to devise the broad framework for sustainable developments for the SIDS. The outcome document called the Mauritius Strategy went even further by claiming that SIDS were already experiencing the vagaries of climate change and that adaptation to these impacts was a major priority. It also underscored the significance of South–South and SIDS–SIDS cooperation in order to strengthen the implementation of the BPOA.[100]

Apart from such efforts, in various COPS, the SIDS made their presence felt well. Before Copenhagen Summit, the Cabinet of Maldives signed a declaration calling for global cuts in carbon emissions to be presented in Copenhagen. The meeting was significant as it took place underwater to send a message to the world that the SIDS are facing climate change driven existential threat and effective steps should be taken at the earliest. At the Copenhagen Summit, the AOSIS proposed a different plan of action against both the Danish Proposal and the draft proposed by the BASIC. They demanded legally binding emission cuts target for the Annex 1 parties in the post Kyoto process also.[101] Moreover, in the Copenhagen Summit the small island states started pressurizing emerging economies to take mandatory emissions cuts for post-2012 period.

At the Cancun Summit of 2010, these states while urging for their protection from the threat of climate change pleaded for establishing a 'global insurance fund'. There were lot of debates regarding the feasibility of such fund. The United Kingdom was not confident about such and favoured the intervention of the UN for investigation.[102] In the CoP 17, Durban, the AOSIS firmly raised their voice that they would disapprove of any proposal to delay any new binding agreement or more ambitious emission reductions until 2020 as these could not safeguard the livelihoods and guarantee the survival of their nations. The SIDS required financial aid from the developed nations to adapt to climate change as they are the victims rather than perpetrators of greenhouse effects.[103] There must be an equity in burden sharing which requires fairness in all aspects of negotiations addressing mitigation, adaptation as well as means of implementation.

The Rio+20 Summit of United Nations Conference on Sustainable Development which was the 20-year review of the Earth Summit reaffirmed that SIDS remain a special case. It agreed to take actions for sustainable development in the SIDS in the coming years. Before this summit the Subregional Preparatory Meeting for SIDS of the Atlantic, Indian Ocean, Mediterranean and South China Sea (AIMS) subregions was held in Mahe, Seychelles on 7 and 8 July 2011.[104] There are some other relevant summits like the Third International Conference of the SIDS that was organized in Samoa in 2014. Before that several interregional preparatory meetings took place like Pacific Regional Preparatory Meeting in Fiji, Caribbean Regional Preparatory Meeting in Jamaica and many more. All of them emphasized that any global deal for taking care of the interests of the SIDS in the face of climate change needs proper capacity building and to mobilize the support from developed countries in the form of technology transfer and aids for development. The 67th session of General Assembly also expressed its expectations from the Third International Conference of the SIDS on account of sustainable development to be promoted there. It emphasized the collaboration between international community and the SIDS.[105]

SIDS are facing multiple problems but the most challenging one is ensuring their capability to adapt to a changing climate as they have limited resource base. The National Adaptation Programme of Action (NAPA) must be mentioned here as it is common across many SIDS. NAPAs provide an UNF-CCC endorsed process for LDCs to identify priorities for addressing the most urgent needs for climate change adaptation by using existing information to present suggestions in an easily conceivable format. It has also been suggested that Island countries like Pacific Island states should start adopting policies to avoid more vulnerability. These might include conditions for settling both the human and infrastructural ones on the higher land, and framing proper plans for evacuation at the time of coastal inundation and many more. Failing to adapt to climate change is detrimental for these island states to a larger extent. If they fail to thrive in a changing climatic condition like sea level rise and storm surges, it would not only alter their physical structures but also would affect their economies, would uproot a large segment of population and would augment resource scarcity, thereby ultimately threatening the security interests of these fragile countries.

3.2. *Response to Increased Intensity of Natural Disasters*

Along with slow-onset process of climate disasters, the rapid-onset weather events like natural disasters have also affected the lives, livelihoods and property of regions across the world. Providing relief to countries affected by natural calamities has always been treated as an important response of global community.

In recent years, the need for early warning systems for natural disasters is gaining ground as governments have realized that better preparedness for ongoing climate shocks will establish capacities for better management of future catastrophes. Over the past several decades, the United Nations is also engaged in the efforts to reduce disaster risk. The decade of 1990s was declared as the international decade for reduction of natural disaster. The first World Conference on Natural Disaster Reduction took place in Yokohoma in 1994 that adopted the Yokohama Strategy and Plan of Action for a Safer World. By recognising the impending losses due to natural disasters, its aim was to develop capacity building institutionally and by human efforts and strengthening infrastructure that would support rebuilding the post disaster-ravaged scenario. The second World Conference was held in Kobe, Hyogo and Japan in early 2005 where the International Strategy for Disaster Reduction was endorsed and the Hyogo Framework for Action (HFA) had been adopted. The required sector-wise work programme to reduce disaster had been first planned. Subsequently, concerns raised about the implications of climate change for disaster risk management led to the agreement between the IPCC and the UN International Strategy for Disaster Reduction to undertake a special report on 'Managing the Risks of Extreme Weather and Disasters' for advancing climate change adaptation.[106] In 2015, the third World Conference on Disaster Risk Reduction has been scheduled with the expectation to devise and to review implementation of the HFA and to adopt a framework for disaster risk reduction for the years following the year 2015. The Hyogo Framework for Action (2005–2015) has been endorsed in the global development frameworks adopted in 2015 and 2016. During this period, a comprehensive global framework has been undertaken, which consisted of some interrelated agreements like the Sendai Framework for Disaster Risk Reduction, the Paris Agreement on Climate Change of 2015 and the Sustainable Development Goals (SDGs). All of these efforts marked the need to make a balanced nexus between sustainable development and disaster risk resilience.

In the face of increasing frequency of disaster risks, the developing countries are concerned more about the needed adaptation measures to the effects of climate change that is felt greatly by the developing countries. The provision of a funding system for their 'loss and damage' had been incorporated in the outcome pact of the Warsaw Climate Summit.[107] Moreover, as the metropolitan cities of developed North are equally vulnerable to natural disasters, there are growing recognitions of the need to build resilient cities. Financial measures like insurance and catastrophe bonds have also become prominent. But they have limited applications. They are out of the reach of the most vulnerable sections of global South who are facing disasters frequently. In the Glasgow Summit of 2021, the demand for funding for loss and damage had been resurfaced, but finally it was omitted from the final draft which created dissatisfaction among many countries.

Thus countries are required to take effective actions to minimize the threat emanating from climate change. A major consequence because of these changes in environment is the displacement of a large number of people which can also be termed a coping strategy against this threat. However, migration is a complex issue with varied dimensions and as it is difficult to identify migrants induced directly by climate change, there are no such instruments that particularly address the climate-induced displacements. In fact migration in general lacks a coherent policy response encompassing all aspects of both internal and international displacements. The next section will analyse some of the efforts and state as well as region-based policy understandings affecting climate migrants, specifically addressing the question of individual rights, state responsibility towards them and identifying adaptation strategies.

4. Global Response to Environmental Change Induced Displacements

Environmental migration is a controversial concept. Therefore, it is difficult at the first place to establish clear definitions and standard terminology of the phenomena. Although International Organization for Migration (IOM) has tried to offer a definition, policymakers and governments across the world failed to develop required laws and safeguards for the plight of environmentally triggered migrants. However, different report programmes and policy papers started mentioning the significance of the issue since the 1990s (see Table 3.3).

Despite the global community has identified the grim scenario that has been developed due to climate migration, the landmark conventions and protocols pertaining to climate change hardly mention methodologies or provisions to address this problem.[108]

The definition of climate migration as propounded by the IOM incorporates both the climate-driven internal and cross-border movements of people. Both of them require some sort of global governance. But efforts to mention of roles and responsibilities of several national and international institutions considering climate migration and their implementation are lukewarm. Additionally, internal migration is considered a matter of respective state's responsibility. However, those migrating exclusively for climate-related problems, like other migrants, are permitted to enjoy all human rights as mentioned in the Universal Declaration of Human Rights (UDHR), International Covenant on Civil and Political Rights(ICCPR) and International Covenant on Economic, Social and Cultural Rights (ICESCR).

Table 3.3 Official Recognitions of Climate Migrations

1985	• Official Derivation of climate refugee in the UNEP report by El-Hinnawi
1990	• The IPCC recognized the role of climate change in inducing human uprooting
2002	• In order to address the climate change issues and the associated requirements for reducing and managing disasters, the International Federation for the Red Cross formed a climate change centre. It focused mainly on the plight of vulnerable people
2005	• A resolution had been adopted by the UN Sub-Commission on the Promotion and Protection of Human Rights. It focuses on the legal implications associated with vanishing territories that had been created because of environmental reasons and the human rights issues of their inhabitants including Indigenous communities
2007	• Climate change induced migration had been first found expression in Antonio Guterres' concern at the Executive Committee meeting in 2007. He was the then High Commissioner of UNHCRs
2008	• A report on climate-induced displacement had been compiled by the Council of Europe Parliamentary Assembly's Committee on Migration, Refugees and Population
2009	• A report on existing and future climate refugees had been issued by Kofi Annan, in the official capacity of the President of the Global Humanitarian Forum
	• The UN General Assembly adopted a resolution on 'Climate Change and Its Possible Security Implications'. On request of the UNGA, the Security Council provided report containing short description of migration and population displacement

Source: Data Collated from Jane McAdam, "Climate Change Displacement and International Law: Complementary Protection Standards", Legal and Protection Policy Series, UNHCR, Division of International Protection, May 2011. Available at http://www.unhcr.org/4dff16e99.ppdf Accessed on June 18, 2014.

However, there are lack of rules and procedures to protect and assist internally displaced persons due to climate change. Mention must be made here about the exception of the Africa Union Convention for the Protection and Assistance of Internally Displaced Persons in Africa. Article 4 of the AU convention identifies the climate refugees and delegated responsibilities to the state parties for their plight.[109]

Special policies were outlined by some countries to take care of cross-border environmentally displaced persons so that they can receive protection within a legal framework and remain at least temporarily without fear of deportation. The United States have enacted such legislation in 1990, whereby giving temporary protected status (TPS) to persons who were unable to return to the homeland because of an ongoing natural disaster.[110] In 2001, such TPS was granted to people from El Salvador who could not get back to their native place for ongoing earthquakes.[111] The European Union has also made effort to provide temporary protection with reference to disaster displacements.[112] At the national level, asylum laws of some countries contain provisions for climate refugees. There is provision in Swedish Asylum Law to extend protection to such people who were not in the condition to get back to their native places. In 2010, Argentina adopted a similar immigration legislation that included provisional residential access to such people.[113] There are some other legislations as well concerning climate refugees. Resolution adopted by the Belgian Senate in 2006, expressing Belgium' concern to fight for these people in the UN. In 2007, a proposal had been made by the Australian Labour Party to build the Pacific Rim coalition for receiving climate refugees.[114] However, in the absence of such legislations, often ad hoc humanitarian responses to these problems are present which are situation specific.

Despite such efforts, it can be said that the institutional roles and responsibilities both at the national and international levels are not sufficient to address the problem of climate migration. The UNHCR which was created for refugee protection has a narrow mandate regarding protection of those displaced internationally by climate change. Since 2007 with the speech of High Commissioner António Guterres, the issue of climate refugee started receiving some recognitions. The first policy paper describing the nexus between climate change and migration had been published by UNHCR in 2008. The 2009 policy paper of UNHCR endorsed the responsibility of its staff to devise strategies for combatting the ills associated with climate-induced displacements.[115] It is thus clear from these developments that there exists a visible change in UNHCR's role over the years. Its role during the Asian Tsunami of 2004 and earthquake in Haiti in 2010 was an expression of this role. There are other agencies as well like the Norwegian Refugee Council, the International Federation of the Red Cross and the International Organization of Migration (IOM) who are dealing with the issue of environmental migration. In various CoPs, climate migration has also become a matter of concern. For instance, the Cancun agreement can be taken into consideration. In Paragraph 14f of the same, such displacements were underlined.[116] Recent summits like the COP21 Paris Agreement specifically identify environmental migration and asking countries to promote their rights.[117] Thus, at the national, global and institutional levels, the issue of migration and climate change are considered significant for policy dialogue. The situation demands that climate migrants need to be incorporated as a new category of refugee who also require protection. Therefore, there exists the need to extend the scope of the Geneva Refugee Convention which did not mention any safeguards specifically for these persons.

5. Summary

Today, people have realized that the intensity and non-linear pattern of climate change and environmental degradation are much more alarming than previously thought but they failed to convert these understandings into meaningful actions. The followings are the roadblocks inhibiting nations to reach an all-encompassing solution to the problem.

- Global parleys on environment reveal that countries' actions are guided by their respective national interests. Both the developed and developing countries are preoccupied with their economic growth and development more than the issue of global warming and climate change. In this eco-politics, nothing fruitful has come out as policymakers pay little attention to the core problem and are engaged in bargaining over sharing the responsibilities and the carbon space. The situation has further been exacerbated due to the lack of an all-pervasive integrated climate change regime which is replaced by a 'climate change regime complex'.[118] Robert O. Keohane and David G. Victor have put it candidly – 'The international institutions that regulate issues related to climate change are diverse in membership and content. They have been created in a context of diverse interests, high uncertainty, and shifting linkages . . . They form a regime complex that is not hardly bound.'[119] Such situation makes the problem of climate change more unmanageable;
- Scarcity of natural resources is related to environmental change and is increasingly a factor contributing to political conflict. This consciousness begets global concerns towards this problem. But as the scarcity occurs differently and at different pace in different countries the priorities also differ that results in delayed response to the problem of environmental decay. In case of food scarcity, mostly developing countries including the small island ones are geographically more fragile to the impacts of climate change induced food shortages and this fragility gets escalated due to widespread poverty induced low adaptive capabilities. Basically, the developing countries are fearing the 'double exposure' to climate change induced food scarcity and economic globalization as food security is related not only to production of food which is

determined by climate variability but also by social and economic policies. With trade liberalization, large transnational agribusiness companies promote more cash crops at the expense of locally produced staples and are spreading their tentacles to monopolize seeds supply in developing countries, thereby increasing the cost of food and affecting native seeds and community-based food system. As the availability of food, access to food and ability to produce food are related to these factors, so the ill-effects of climate change on agriculture cannot be mitigated in isolation from these issues.

The supply, quality and distribution of water are affected by environmental change as well. But it is also true that the access and availability of water are restricted because of the commodification of this resource with economic globalization. As a result, the price of this public good is controlled by the rules of the market.[120] It implies private control over a life-sustaining environmental resource. The marketization and privatization of water sometimes have made the resource scarce to vulnerable sections despite its apparent abundance. Besides during climate-induced drought, the cost of water may increase due to decrease in water supply which inevitably push the poorer section into a more water-stressed situation. In climate negotiations, this nexus between water resource and climate change privatization of water is hardly discussed; therefore, the problems related to water governance has remained unsolved.

Forest is another important resource that is affected by climate change and if properly maintained can be used as a means of climate change mitigation. The global community responded to each of the challenges associated with deforestation both collectively and sometimes national interests of different forested countries have guided their actions. In the REDD+ negotiations, the world has witnessed many such coalitions and individual actions as the parties are concerned for forests for various reasons. The G77 collective is therefore fractured, and there were many shades of differentiations related to forest issues within it. While Brazil was vocal about protecting its sovereign rights over its part of Amazonian forests, Costa Rica, Bolivia as part of CfRN were leaders in forest conservation and they are significant actors in the REDD+ process. Apart from Costa Rica and Bolivia, other Latin American countries like Honduras, Paraguay, Panama and Peru, at times joined by Argentina, Mexico and Ecuador, championed a 'nested approach' arguing that it was the most effective way to combine national and sub-national methods of accounting and crediting. Such an approach would support a national-level accounting system while allowing sub-national projects to earn credits.[121]

Both Brazil and Indonesia are REDD+ powerhouses, but they maintained their individual positions in the negotiations instead of joining any leading forest coalitions like CfRN although some of their concerns converge. Basically, they differ in their respective ways of mitigating the problem. For instance, despite many forest-related coalitions like

COMIFAC and CfRN favoured market mechanisms in the REDD+ process, Brazil strictly denied that and instead it had proposed that developed countries should establish and replenish an international fund to support REDD+ projects and developing countries should receive credits for emission reductions (exchangeable for funds).[122] Similarly, Indonesia which also not resorted to join forest coalitions raised its own voice underscoring the capacity building measures to implement projects under REDD+. Apart from this, regional groupings like ASEAN were also addressing the problem of deforestation and trade of illegal logging. ASEAN summits and the ASEAN Ministers of Agriculture and Forestry (AMAF) regularly address these issues. Such region-specific initiatives and divisive political interests have impeded the success of these efforts. So the effort for proper conservation, management and sustainable development of forest lacks global consensus and is often guided by self-interested behaviours.

The world's reliance on fossil fuel based economic growth has also contributed to climate change. The concern for this threat therefore is bound to change the landscape of energy policy across the world. But each effort has its own shortcomings.

- Natural gas can be used as a substitute for coal-fired electricity generation, but countries may become vulnerable through their gas import. Because political animosity often may result in sanctions on trade. For example, Russia cut gas exports to Europe in the context of pricing dispute with Ukraine through which approximately 80 per cent of Russian gas passes on its way to Europe.
- The CCS technology which is proposed to deploy across power sectors is costly and requires some more energy which is called 'energy penalty' for carbon capture and compression.
- Renewables are often touted as essential option for combating climate change but countries need to depend on importing technologies and materials necessary for renewable power and storage.
- Biofuels are also not beyond criticism as their production requires large amount of natural gas which may increase energy insecurity in places which depend on natural gas import. Its production directly or indirectly may lead to land use change resulting in emissions that massively overwhelm any emissions reductions that biofuels production leads. It has also clash with food production. Though second-generation biofuels, that is, biofuels produced from waste products, or crops grown on marginal land are free from these issues,[123] these methods are yet to be substantiated.[124]
- Nuclear power though is one of the near-zero carbon sources is not used widely because of the leakage and sabotage dangers associated with it, and it requires Uranium on which commercial reactors depend for fuel. So it gives birth to another type of energy security vulnerability.

However, this is not the only problem. Reluctance on the part of the countries to materialize the efforts to switch over to low-carbon-intensive energy sources is also responsible for delayed response. Not only that, due to uneven resource concentration like oil in the OPEC countries, the problem gets aggravated.

- Extreme weather conditions are hitting the world more vigorously than before. But while the developed countries can handle these crises more efficiently due to financial and technological affluence, in the developing ones poverty limits adaptation capabilities. In such a situation, they depend largely on public financial support from developed countries. For the poor vulnerable regions, adaptation and resilience to climate change are more significant than the mitigation efforts to climate change.
- Climate migration as a consequence as well as a coping strategy is also a critical one. There is no legal basis for defining these types of migrants under International Law. Though they have received the aid and are treated equally like other migrants, climate measures lack the ability to look after the special conditions of them. Sometimes development works uproot people from their original habitat and has created ecological distress. These issues are not taken care of by any measures related to stop environmental degradation.

So global response to environmental degradation and climate change fails to reach a solution that is all encompassing. The climate vulnerability is associated with many other stressors, and countries are mostly unaware of them. Rising pace of emissions is only one facet of the problem. In various climate summits mainly from different quarters, agitation comes to the forefront regarding the sharing of burden. All the stakeholders are fearing that the imposition of constraints on their rising emissions would ultimately put them at a disadvantageous state both politically and economically. Climate-induced food shortages, water scarcity, degradation of biodiversity and frequency of extreme weather events have affected all the countries but as the economic situations and capacity to stand against these threats differ and as there are multiple layers of uncertainty involved in addressing the problem and planning, there cannot be a global solution to this global problem of environmental decay. Narrow politics, self-interest and profit-making intentions have impeded the success of a global climate regime. Given such a state of disarray and confusion in global environmental talks, countries are therefore required to first identify and realize the density and various dimensions of the threat and then cooperate among themselves so that they can negotiate rationally to translate the commitments into tangible actions.

Notes

1 UNEP, *Integrating Environment and Development: 1972–2002*, p.4. Available at www.unep.org/GEO/geo3/pdfs/Chapter1.pdf Accessed on March 14, 2013.
2 The most notable moves in the run up to the Rio Summit were the setting up of World Commission for Environment and Development (Brundtland commission)

by UN General Assembly in 1983 and the signing of Montreal Protocol in 1987 to address the problem of Ozone depletion.

3 Paul G. Harris, "Climate Change", in Gabriela Kutting (ed.), *Global Environmental Politics: Concepts, Theories and Case Studies*, Routledge: London, 2011, p.110.

4 Agenda 21 is an international blueprint outlining the required actions for governments, international organisations, industries and the community to achieve sustainability. Its objective was to alleviate poverty, hunger, sickness and illiteracy while halting the deterioration of ecosystem which sustains life.

5 The act recognised that the animal and plant life on the earth is endangered by excessive emission. It was subscribed by more than hundred countries from both the developed and developing worlds except the United States.

6 UNFCCC, Report of the Conference of Parties on its Third Session, held at Kyoto from December 1–11, 1998, p.9. Available at https://cdm.unfccc.int/Reference/COPMOP/08a01.ppdf Accessed on March 14, 2013.

7 Andrew Hurrell and Sandeep Sengupta, "Emerging Powers, North-South Relations and Global Climate Politics", *International Affairs*, Vol.88, No.3, 2012, p.471.

8 Jayanta Basu, "Climate Emergency CoP 25: Loss and Damage 'Fighting Out' in Madrid", *Down to Earth*, December 13, 2019. Available at downtoearth.org.in Accessed on January 23, 2022.

9 Montek S. Ahluwalia and Patel Utkarsh, "The Glasgow Summit on Climate Change: What Has It Achieved? ", *Live Mint*, November 14, 2021. Available at livemint.com Accessed on January 18, 2022.

10 Timperley Jocelyn, "The Broken $100-Billion Promise of Climate Finance – and How to Fix It", *Nature*, October 20, 2021. Available at https://www.nature.com/articles/d41586-021-02846-3 Accessed on January 23, 2022.

11 Dubrin Adam and Bowden George, "COP26: UK Pledges £290m to Help Poorer Countries Cope with Climate Change", *BBC News*, November 8, 2021. Available at https://www.bbc.co.uk/news/uk-59202129?at_campaign=KARANGA&at_medium=RSS Accessed on January 19, 2022.

12 Timperley Jocelyn, No.10.

13 Damian Carrington, "What Is 'Loss and Damage' and Why Is It Critical for Success at Cop26?", *The Guardian*, November 13, 2021. Available at https://www.theguardian.com/environment/2021/nov/13/what-is-loss-and-damage-and-why-is-it-critical-for-success-at-cop26 Accessed on January 23, 2022.

14 Urs Steiner Brandta and Gert Tinggaard Svendsen, "Hot Air in Kyoto, Cold Air in The Hague – The Failure of Global Climate Negotiations", *Energy Policy*, No.30, 2002, pp.1191–1192.

15 It implies that the granted quota of permits is higher than actual emission. The presence of it means that nations do not have to undertake real reductions when actual emissions are already lower than 1990 levels.

16 Purusottam Bhattacharya, "Quest for Environmental Security in Europe: A Case Study of the Common Environmental Policy of the European Union", in Purusottam Bhattacharya and Hazra Sugata (eds.), *Environment and Human Security*, Lancer Books: New Delhi, 2003, pp.141–142.

17 Uttam Kumar Sinha, "Climate Summit at Copenhagen: Negotiating the Intractable", *Strategic Analysis*, Vol.33, No.6, November 2009, p.797.

18 Chandrasekhar Dasgupta, "The Future of Global Climate Regime", *Foreign Affairs Journal*, Vol.7, No.3, July-September 2012, p.276.

19 European Commission, *EU at COP 26 Climate Change Conference*. Available at https://ec.europa.eu/info/strategy/priorities-2019-2024/european-green-deal/climate-action-and-green-deal/eu-cop26-climate-change-conference_en Accessed on December 13, 2021.

20 Gørild M. Heggelund and Fritzen Buan Inga, "China in the Asia-Pacific Partnership: Consequences for UN Climate Change Mitigation Efforts?", *International Environmental Agreements: Politics, Law and Economics*, Vol.9, No.3, 2009, p.309.

21 Parkash Chander, "How to Talk Climate Change in Paris", *The Hindu*, December 17, 2014.

22 Anmar Frangoul, "China's Shock Climate Deal with the U.S. Sparks Some Cautious Optimism", *CNBC*, November 11, 2021. Available at COP26: U.S.-China declaration on climate welcomed (cnbc.com) Accessed on January 23, 2022.

23 Mizan R. Khan, "From Cancun to Durban: Is There Any Likelihood of a New Climate Regime?", *BIIS Journal*, Vol.32, No.1, January 2011, p.55.

24 Praful Bidwai, "Durban: Road to Nowhere", *Economic and Political Weekly*, Vol.XLVI, No.53, December 31, 2011, pp.10–11.

25 T. Jayaraman, "*India and Climate Talks Imperatives*", *The Hindu*, November 18, 2013.

26 Bruce A. McCarl, Mario A. Fernandez, Jason P.H. Jones and Marta Wlodarz, "Climate Change and Food Security", *Current History*, Vol.112, No.750, January 2013, pp.36–37.

27 Conference Statement, *The Changing Atmosphere: Implications for Global Security*, Conference held in Toronto Canada, June 27–30, 1988, pp.293–301. Available at http://www.cmos.ca/ChangingAtmosphere1988e.pdf Accessed on May 15, 2014.

28 UNFCCC text. Available at http://unfccc.int/essential_background/convention/background/items/1353.php Accessed on May 16, 2014.

29 UNFCCC, Technical Paper, *Challenges and Opportunities for Mitigation in the Agricultural Sector*, November 21, 2008, p.11. Available at http://unfccc.int/resource/docs/2008/tp/08.pdf Accessed on May 16, 2014.

30 UNFCCC, *Land Use, Land-Use Change and Forestry (LULUCF)*. Available at https://unfccc.int/methods/lulucf/items/3060.php Accessed on May 16, 2014.

31 Eva Wollenberg, Marja-Liisa, Tapio Bistrom and Maryanne Grieg-Gran, "Climate Change Mitigation and Agriculture: Designing Projects and Policies for Smallholder Farmers", in Eva Wollenberg et al. (eds.), *Climate Change Mitigation and Agriculture*, Earthscan: London, 2012, pp.11–12.

32 Ibid., p.12.

33 George Wamukoya, *COP 15 Outcomes for Reducing Emissions from Agriculture and Other Land Uses*, 2010. Available at http://www.iisd.org/pdf/2010/03_REDD_II_Hue_Agriculture.pdf Accessed on May 19, 2014.

34 Gender CC-Women for Climate Justice, *Briefing Paper on UNFCCC and Agriculture*, p.5. Available at www.gendercc.net/uploads/media/Briefing_Paper_Agriculture_2012.pdf Accessed on May 19, 2014.

35 Ibid.

36 FAO, *A Guide to Agriculture at UNFCCC COP 19*, November 2013, pp.1–3. Available at http://www.fao.org/docrep/019/ar716e/ar716e.pdf

37 Gender CC-Women for Climate Justice, No.34, p.8.

38 Indrajit Bose, Arnab Pratim Dutta and Souparno Banerjee, "Frozen at Gateway", *Down to Earth*, December 16–31, 2012, p.42, p.35.

39 Gender CC-Women for Climate Justice, No.34, pp.8–9.

40 Eva Wollenberg, Marja-Liisa, Tapio Bistrom and Maryanne Grieg-Gran, No.31, p.13.

41 Cheikh Mbow, Pete Smith, David Skole, Lalisa Duguma and Mercedes Bustamante, "Achieving Mitigation and Adaptation to Climate Change Through Sustainable Agroforestry Practices in Africa", *Current Opinion in Environmental Sustainability*, 2014, p.8. Available at www.sciencedirect.com Accessed on May 19, 2014.

42 Emily K. Anderson and Hisham Zerriffi, "Seeing the Trees for the Carbon: Agroforestry for Development and Carbon Mitigation", *Climate Change*, Vol.115, No.3–4, December 2012, p.742.

43 Agenda 21, Chapter 18. *Protection of the Quality and Supply of Freshwater Resources: Application of Integrated Approaches to the Development, Management and Use of Water Resources*, 1992. Available at http://www.earth summit2002.org/ic/freshwater/reschapt18b.html Accessed on May 27, 2014.

44 Claudia Sadoff and Mike Muller, *Water Management, Water Security and Climate Change Adaptation: Early Impacts and Essential Responses*, Global Water Partnership Technical Committee Background Paper No.14. Available at www.gwp.org/Global/GWP...Files/.../tec14.pdf Accessed on May 27, 2014.

45 UN-Water Thematic Initiative, *Coping with Water Scarcity*, August 2006, p.4. Available at www.un.org/waterforlifedecade/.../2006_unwater_coping_with_water_sc Accessed on May 28, 2014.

46 J.P. Msangi, "General Introduction", in J.P. Msangi (ed.), *Combating Water Scarcity in Southern Africa: Case Studies from Namibia*, Springer: Dordrecht, 2014, pp.1–2.

47 Global Water Partnership, *A Handbook for Integrated Water Resources Management in Basins*, p.2. Available at http://www.gwp.org/Global/ToolBox/References/A%20Handbook%20for%20Integrated%20Water%20Resources%20Management%20in%20Basins%20%28INBO,%20GWP,%202009%29%20ENGLISH.pdf Accessed on May 28, 2014.

48 UNEP, "Summary of the First International Environment Forum for Basin Organizations", *Basin Organizations Forum Bulletin*, Vol.227, No.1, 2014, p.5. Available at http://www.unep.org/delc/Portals/119/ForumBasinOrganization/IISD-basinforum-report.pdf Accessed on May 27, 2014.

49 David Humphreys, "International Forest Politics", in Gabriela Kutting (ed.), *Global Environmental Politics: Concepts, Theories and Case studies*, Routledge: London, 2011, pp.136–139.

50 United Nations Forum for Forests, *IPF/Iff Process (1995–2000)*. Available at http://www.un.org/esa/forests/ipf_iff.html Accessed on June 5, 2014.

51 P. Moutinho, M. Santilli, S. Schwartzman and L. Rodrigues, *Why Ignore Tropical Deforestation? A Proposal for Including Forest Conservation in the Kyoto Protocol*. Available at www.fao.org/docrep/009/a0413e/a0413e06.htm Accessed on June 7, 2014.

52 Jen Iris Allan and Peter Dauvergne, "The Global South in Environmental Negotiations: The Politics of Coalitions in Redd+", *Third World Quarterly*, Vol.34, No.8, 2013, p.1313.

53 Submission by the Governments of Papua New Guinea & Costa Rica, Eleventh Conference of the Parties to the UNFCCC: Agenda Item 6, *Reducing Emissions from Deforestation in Developing Countries: Approaches to Stimulate Action*. Available at rainforestcoalition.org/.../COP-11AgendaItem6-Misc.Doc.FINAL.pdf Accessed on June 10, 2014.

54 UNFCCC, *Fact Sheet: Reducing Emissions from Deforestation in Developing Countries: Approaches to Stimulate Action*, June 2009, p.1. Available at unfccc.int/.../backgrounders/.../fact_sheet_reducing_emissions_from_def... Accessed on June 6, 2014.

55 Center for Climate and Energy Solution, COP 12 Report, *Twelfth Session of the Conference of the Parties to the UN Framework Convention on Climate Change and Second Meeting of the Parties to the Kyoto Protocol*. Available at www.c2es.org/international/negotiations/cop-12/summary Accessed on June 6, 2014.

56 Jen Iris Allan and Peter Dauvergne, No.52, p.1314.

57 UNFCCC, No.54, p.2.

58 PEW Center on Global Climate Change, *Summary: Copenhagen Climate Summit*, 2009. Available at http://www.c2es.org/international/negotiations/cop-15/summary Accessed on June 6, 2014.

59 "One of the elements countries need to develop to participate in REDD+ is a Forest Reference Emission Level and/or Forest Reference Level (FREL/FRL). The UNFCCC has defined FREL/FRLs as benchmarks for assessing each country's performance in implementing REDD+ activities. In UNFCCC COP decisions the term forest reference emission levels and/or forest reference levels (FREL/FRLs) is used. Though the UNFCCC does not explicitly specify the difference between a FREL and a FRL, the most common understanding is that a FREL includes only emissions from deforestation and degradation, where as a FRL includes both emissions by sources and removals by sinks, thus it includes also enhancement of forest carbon stocks". FAO, UNDP, UNEP, *UN-REDD Programme: Emerging approaches to Forest Reference Emission Levels and/or Forest Reference Levels for REDD+*, October 2014, pp.4–8. Available at https://unfccc.int/files/land_use_and_climate_change/redd_web_platform/application/pdf/redd_20141113_unredd_frel.pdf Accessed on February 8, 2015.

60 "COP16 Cancun, 2010: In Which Poor Countries Gave In", *Down to Earth*, January 1–15, 2011, p.34.

61 Chad Carpenter, *Taking Stock of Durban: Review of Key Outcomes and the Road Ahead*, UNDP Environment and Energy Group, April 2012, pp.27–28. Available at http://www.undpcc.org/docs/Bali%20Road%20Map/English/UNDP_Taking%20Stock%20of%20Durban.pdf Accessed on June 7, 2014.

62 Ibid.

63 Uthra Radhakrishnan, "Power Games at UN Climate Talks", *Down to Earth*, December 15, 2013, p.2. Available at http://www.downtoearth.org.in/content/power-games-un-climate-talks Accessed on March 21, 2015.

64 UNFCCC, *What Is REDD+?* Available at https://unfccc.int/topics/land-use/workstreams/redd/what-is-redd Accessed on January 19, 2022.

65 "Protect Papua New Guinea's Rainforest: COP26 REDD+ Carbon Credits Auction", *Environmental Finance*, October 18, 2021. Available at environmentalfinance.com Accessed on January 20, 2022.

66 David Humphreys, "International Forest Politics", in Gabriela Kutting (ed.), *Global Environmental Politics: Concepts, Theories and Case Studies*, Routledge: London, 2011, pp.143–144.

67 Georgina Rannard and Francesca Gillett, "COP26: World Leaders Promise to End Deforestation by 2030", *BBC News*, November 2, 2021. Available at https://www.bbc.com/news/science-environment-59088498 Accessed on January 20, 2022.

68 *Kimberley Declaration*, 2002. Available at www.tebtebba.org/.../17-rio-10-world-summit-on-sustainable-d... Accessed on June 10, 2014.

69 Tom Griffiths, "Seeing 'RED'?: 'Avoided Deforestation' and the Rights of Indigenous Peoples and Local Communities", *Forest People's Programme*, June 2007, p.17. Available at http://www.forestpeoples.org/sites/fpp/files/publication/2010/01/avoideddeforestationredjun07eng_0.pdf Accessed on February 4, 2015.

70 Two *Very Different Views* on the Warsaw REDD *Deal* from Indigenous Peoples Organisations, 2013. Available at http://www.redd-monitor.org/2013/12/07/two-very-different-views-on-the-warsaw-redd-deal-from-indigenous-peoples-organisations/ Accessed on June 23, 2014.

71 Georgina Rannard and Francesca Gillett, No.67.

72 Mikael Hook and Xu Tang, "Depletion of Fossil Fuels and Anthropogenic Climate Change – A Review", *Energy Policy*, Vol.52, January 2013, p.801.

73 Carbon Capture and Storage Association, *What Is CCS?* Available at http://www.ccsassociation.org/ Accessed on March 8, 2015.

74 Antony Froggatt and Michael A. Levi, "Climate and Energy Security Policies and Measures: Synergies and Conflicts", *International Affairs*, Vol.85, No.6, 2009, p.1134.

75 Ference L. Toth, "Nuclear Power as a possible Response to Climate Change", in Artur Gradziuk and Ernest Wyciszkiewicz (eds.), *Energy Security and Climate Change: Double Challenge for Policy Makers*, The Polish Institute of International Affairs: Warsaw, pp.73–94, p.81.

76 Agenda 21, Chapter 14, *Promoting Sustainable Agriculture and Rural development*. Available at http://www.fao.org/sd/erp/toolkit/Books/SARDLEARN-ING/CD-SL/Sources/Agenda%2021-chapter%2014.htm Accessed on June 21, 2014.

77 The World Bank, *Fuel for Thought: An Environmental Strategy for the Energy Sector*, June 2000, pp.37–38. Available at http://documents.worldbank.org/curated/en/2000/06/443544/fuel-thought-environmental-strategy-energy-sector Accessed on June 21, 2014.

78 Susan R. Fletcher and Larry Parker, *Climate Change: The Kyoto Protocol, Bali Action Plan, and International Actions*, CRS Report for Congress, January 2008. Available at http://www.house.gov/sites/members/nc04_price/issues/uploadedfiles/climate5.pdf Accessed on June 23, 2014.

79 *Gleneagles Plan of Action on Climate Change, Clean Energy and Sustainable Development*, July 2005. Available at https://www.gov.uk/government/uploads/system/uploads/attachment_data/file/48584/gleneagles-planofaction.pdf Accessed on June 24, 2014.

80 IEA, *World Energy Outlook*, 2006, p.36. Available at http://www.worldenergyoutlook.org/media/weowebsite/2008–1994/WEO2006.pdf Accessed on June 23, 2014.

81 Quoted in Bernhard May, "Energy Security and Climate Change: Global Challenges and National Responsibilities", *South Asian Survey*, Vol.17, No.1, 2010, p.21.

82 Navroz K. Dubash and Ann Florini, "Mapping Global Energy Governance", *Global Policy*, Vol.2, September 2011, p.10.

83 *Final Chairman's Summary: First Major Economies Meeting on Energy Security and Climate Change*, White House Council on Environmental Quality, September 27–28, 2007. Available at http://2001-2009.state.gov/g/oes/climate/mem/93021.htm Accessed on June 24, 2014.

84 International Energy Agency, *Progress with Implementing Energy Efficiency Policies in the G 8*, 2009. Available at http://www.iea.org/publications/freepublications/publication/G8Energyefficiencyprogressreport.pdf Accessed on June 24, 2014.

85 Artur Gradziuk, "Energy Security and Climate Change: Seeking a Balance in New Reality", in Artur Gradziuk and Ernest Wyciszkiewicz (eds.), *Energy Security and Climate Change: Double Challenge for Policy Makers*, The Polish Institute of International Affairs: Warsaw, 2009, p.22.

86 Froggatt Antony, *China, EU and US Cooperation on Climate and Energy*, Research Paper, Chatham House, March 29, 2021. Available at International Affairs Think Tank Accessed on January 20, 2021.

87 Ibid., pp.21–22.

88 UNFCCC, *Global Energy Interconnection Is Crucial for Paris Goals*, November 28, 2021. Available at https://unfccc.int/news/global-energy-interconnection-is-crucial-for-paris-goals Accessed on January 20, 2022.

89 Kandekh K. Yumkella, "Sustainable Energy for All: Towards Rio+20", *UN Chronicle*, No.1 and 2, 2012, p.21.

90 UN News Service, '*The Future Starts Now,*' *Ban Says at Launch of UN Decade of Sustainable Energy for All*, June 5, 2014. Available at http://www.un.org/apps/news/story.asp?NewsID=47969#.U6z7ytdDu8Y Accessed on June 27, 2014.

91 Navroz K. Dubash and Ann Florini, No.82, p.11.

92 Antony Froggatt and Michael A. Levi, No.74, p.1134.

93 Ibid., p.1135.

94 Marlyn A. Brown and Michael Dworkin, "The Environmental Dimension of Energy Security", in Benjamin K. Sovacool (ed.), *The Routledge Handbook of Energy Security,* Routledge: London, pp.185–187.

95 Shelagh Whitley, *Time to Change the Game: Fossil Fuel Subsidies and Climate*, Overseas Development Institute, November 2013, pp.1–21. Available at www.odi.org.uk/sites/odi.../8669.pdf Accessed on June 25, 2014.

96 UNEP, *Emerging Issues for Small Island Developing States: Results of the UNEP Foresight Process*, p.41. Available at http://www.indiaenvironmentportal.org.in/files/file/Emerging%20issues%20for%20small%20island %20developing%20 states.pdf Accessed on June 12, 2014.

97 Ilan Kelman and Jennifer J. Wes, "Climate Change and Small Island Developing States: A Critical Review", *Ecological and Environmental Anthropology*, Vol.5, No.1, 2009, p.1.

98 *Barbados Programme of Action for the Sustainable Development of Small Island Developing States*, 1994. Available at http://www.unep.ch/regionalseas/partners/sids.htm Accessed on June 12, 2014.

99 "*Nassau Forum Adopts Declaration, Strategy Paper in* Preparation *for Mauritius Meeting on Small Island States*", Press Release, DEV/2456, January 30, 2004. Available at www.un.org/News/Press/docs/2004/dev2456.doc.htm Accessed on June 13, 2014.

100 UN General Assembly, *Mauritius Strategy for the Further Implementation of the Programme of Action for the Sustainable Development of Small Island Developing States*, Report of the Secretary General, October 3, 2005, pp.3–10. Available at http://unctad.org/en/Docs/a60d401_en.pdf Accessed on June 13, 2014.

101 *Proposal by the Alliance of Small Island States (AOSIS) for the Survival of the Kyoto Protocol and a Copenhagen Protocol*, December 11, 2009. Available at http://www.indiaenvironmentportal.org.in/content/293517/proposal-by-the-alliance-of-small-island-states-aosis-for-the-survival-of-the-kyoto-protocol-and-a-copenhagen-protocol/ Accessed on June 14, 2014.

102 Louise Gray, "Cancun Climate Change Summit: Small Island States in Danger of 'extinction' ", *The Telegraph*, December 1, 2010.

103 Clem Tisdell, *Global Warming and the Future of Pacific Island Countries*, Working Paper No.147, The University of Queens Land, 2007, p.16. Available at http://www.uq.edu.au/rsmg/docs/ClemWPapers/EEE/WP147.pdf

104 *Outcome Document of Rio+20 Subregional Preparatory Meeting of SIDS of the Atlantic, Indian Ocean, Mediterranean and South China Sea (AIMS) Subregions*, July 2011. Available at www.uncsd2012.org/.../documents/AIMS%20Rio+20%20Outcome%20d Accessed on June 14, 2014.

105 IISD Reporting Services, "Summary of the First International Environment Forum for Basin Organizations", *Basin Organizations Forum Bulletin*, Vol.227, No.1, 2014, pp.1–13. Available at http://www.iisd.ca/download/pdf/sd/crsvol227num1e.pdf Accessed on March 21, 2015.

106 Madelaine C. Thomson, "Climate Change and Disaster Risk Management: Challenges and Opportunities", in Sarah Boulter, Jean Palutikof, David John Karoly and Daniela Guitart (eds.), *Natural Disasters and Adaptation to Climate Change*, Cambridge University Press: Cambridge, pp.8–13.

107 UNDP, UNEP, UNESCAP, UNFCCC, UNISDR and WMO, *TST Issue Brief: Climate Change and Disaster Risk Reduction*, 2013. Available at http://sustainabledevelopment.un.org/content/documents/2301TST%20Issue%20Brief_CC&DRR_Final_4_Nov_final%20final.pdf Accessed on August 25, 2014.

108 Available at http://unfccc.int/resource/docs/2008/smsn/igo/022.pdf Accessed on June 16, 2014.

109 Susan Martin, "Climate Change, Migration and Governance", *Global Governance*, Vol.16, No.9, January–March 2010, p.402.

110 Ibid., p.407.

111 Jane McAdam, *Climate Change Displacement and International Law: Complementary Protection Standards*, Legal and Protection Policy Series, UNHCR, Division of International Protection, May 2011, p.39. Available at http://www.unhcr.org/4dff16e99.pdf Accessed on June 18, 2014.

112 Vikram Kolmannskog, *Climate Change-Related Displacement and the European Response*, Paper presented at SID Vijverberg Session on Climate Change and Migration, The Hague, January 20, 2009. Available at sideurope.files.wordpress.com/2009/02/presentation-kolmannskog.doc Accessed on June 17, 2014.

113 Jane McAdam, No.111, pp.39–41.

114 Ibid., pp.40–41.

115 Nina Hall, *Climate Change and Institutional Change in UNHCR*, Conference Paper for UNU-EHS Summer Academy on Protecting Environmental Migration: Creating New Policy and Institutional Frameworks, July 25–31, 2010, pp.10–11. Available at http://www.ehs.unu.edu/file/get/5404 Accessed on June 17, 2014.

116 IOM, *Climate Change, Environmental Degradation and Migration*, International Dialogue on Migration, No.18, 2012. Available at www.iom.int/.../workshops/clim.. Accessed on June 17, 2014.

117 Dina Ionesco, *Climate Migration: From the Paris Agreement to the Global Compact for Migration*, IOM, UN Migration Blog, November 30, 2017. Available at https://weblog.iom.int/climate-migration-paris-agreement-global-compact-migration Accessed on January 20, 2022.

118 According to Keohane and David Victor, regime complexes are marked by 'connections between the specific and relatively narrow regimes but the absence of an overall architecture or hierarchy that structures the whole set'. There are three forces which help in explaining the reasons behind such fragmentation like distribution of interests for which no single system combating climate change has emerged; uncertainty about the gains that the nations will accrue and their vulnerability from climate change related regulations and finally linkages through which an issue area can expand in size.

119 Robert O. Keohane and David G. Victor, "The Regime Complex for Climate Change", *Perspectives on Politics*, Vol.9, No.1, March 2011, p.19.

120 Karen L. O'Brien and Robin M. Leichenko, *Climate Change, Globalization and Water Scarcity*, p.6. Available at http://www.google.co.in/url?sa=t&rct=j&q=&esrc=s&source=web&cd=1&ved=0CCEQFjAA&url=http%3A%2F%2Fwww.zaragoza.es%2Fcontenidos%2Fmedioambiente%2FcajaAzul%2F17S6-P2-OBrien ACC.pdf&ei=ZwwCVJffItWVuATi2YLIAw&usg=AFQjCNEbO6WuzpsbMp22SgxgWBKgFY9mmg&bvm=bv.74115972,d.c2E Accessed on August 29, 2014.

121 Jen Iris Allan and Peter Dauvergne, No.52, p.1316.

122 Ibid., p.1319.

123 Antony Froggatt and Michael A. Levi, No.74, pp.1134–1138.

124 Girjesh Pant, "The Future of Energy Security Through a Global Restructuring", *South Asian Survey*, Vol.17, No.1, 2010, p.39.

4 Indian Response to Environmental Degradation and Climate Change

India is a significant player in the global environmental politics. It is a biologically diverse country with many climatic zones. Such geographical landscape along with the environment-sensitive economy is facing challenges due to changes in climate. So, the situation demands effective measures to be taken so that the climate vagaries can be averted. In the negotiation tables in various global climate-related forums, it acts as an active and leading voice instead of being a mere silent spectator. In the domestic arena, also its response to this looming crisis contains many dimensions. Policy responses to environmental degradation here have taken various forms. There are now urges to implement environmental rules that are inherent in the Constitution, to enact various environmental laws that are problem specific, to represent India's environmental consciousness on the global platform and many other relevant moves. Each of these efforts has proven successful on various occasions while each has its own shortcomings.

From the very outset, India has expressed her deep concern for the environment while underpinning the significance of developmental need of the developing world in general and India in specific. This is evident in the speech of Late Mrs. Indira Gandhi, former PM of India, that was delivered nearly four decades ago at the Plenary Session of the United Nations Conference on Human Environment, Stockholm in 1972. The graveness of the problems associated with environmental degradation was recognized by her vigorously when she argued that the climate crisis would change the future goal of the entire planet, and none would remain unaffected irrespective of the circumstance, status and strength.[1] She blamed poverty and need as the greatest polluters and stressed the role of insufficient levels of development in generating environmental problems in developing countries.

1. Landmark Environmental Legislations and the Environmental Provisions in the Constitution of India

The pro-environment speech by Mrs. Gandhi as mentioned earlier basically ushered in a new era of Indian environmentalism – it was the harbinger of marked shifts in the environmental policy fabric of India. A number of

DOI: 10.4324/9781003271192-5

significant steps had been taken by India to materialize the decisions taken at the conference. She therefore introduced new legislations concerning environmental protection, amendments to the Constitution and formation of institutions and boards to implement the enactments. Soon after the conference, the Wildlife Protection Act 1972 was drafted and was enacted. Poaching and illegal trade in wildlife has been banned by this act and its derivatives decoded ways for identifying major forests in India to be declared as Tiger Reserves.[2] The parliament of India had also enacted the Water (Prevention and Control of Pollution) Act 1974, the Air Act 1979, The Forest (Conservation) Act 1980 and many others in the subsequent years which were basically 'recommendatory guidelines'.

After the Stockholm Conference, several amendments to the constitution were made in order to make the principles of the conference effective in the Indian scenario. One of such efforts is visible in the 42nd Amendment to the Indian constitution by virtue of which Article 48 A was inserted in the constitution for developing efforts for the protection and improvement of the environment and to build safeguards the forests and wild lives of the country. Another significant development in the environmental jurisprudence in India was the incorporation of Article 51(g) dealing with Fundamental duties. It casts a duty following Article 48A.[3] In the concurrent list, two entries were also incorporated for the protection of forests and wild lives (Entry 17 A and 17 B, respectively). The judiciary of the country has also played a pivotal role in enforcing provisions in constitution and relevant laws pertaining to environment. In interpreting the Constitution, the Supreme Court of India has facilitated the concerns for environment too. For instance, while interpreting Article 21 of the Constitution, the apex body of the judiciary includes the right to have pollution free air and water.[4]

To build the institutional infrastructure for implementing and reviewing the environmental legislations, several committees came into being. The ND Tiwari committee of 1980 was one of such kind. It first recommended the formation of the Department of environment and also suggested the Environment Impact Assessment (EIA) for the periodical review of the industries. Finally, the Department. of Environment was built by the GoI in 1980 and later was transferred to the MoEF[5] in 1985.[6] However, despite such efforts the ethos of environmental protection and preservation remained confined within the governmental realm. The scenario had been changed with the adoption of Ganga Action Plan (the basis of this plan was a petition filed by eminent Supreme Court constitutional lawyer M. C. Mehta claiming that the river Ganges was being polluted by towns and industries) and the Bhopal Gas Accident that took many lives underscoring the need for a comprehensive environmental legislation. The industrial disaster in India as the Bhopal incident forced to change in the existing legal disaster management framework. Several new laws came into prominence. The judiciary also started entering the environmental domain more fiercely. It enhanced the environmental awareness of the Indian populace to a great extent.[7] All of these developments contributed towards the comprehensive legislation of the Environment (Protection) Act 1986. With this act, the Central government acquired permission for adopting measures to improve and protect the environment.[8] The Central Pollution Control Board (CPCB) had been established as well, and it enabled the formation of Pollution Control Boards at the state levels which would act under the overall control and supervision of CPCB. All of these laws and other events made our population aware of the fact that without necessary actions, environmental degradation, ecological imbalances and climate change might imperil our existence.

2. India and the UN Response to Environmental Change

India is an extremely influential actor in the global environmental politics. She has raised her firm voice in various landmark environmental summits in favour of environmental protection without jeopardizing her national interest. In various climate summits under the aegis of the UN, her national interest often clashes with the process and outcomes, but that did not mould her climate posture.

2.1. India's Role in the International Environmental Regime and the Corresponding Domestic Actions

2.1.1. The UNFCCC Process and India

In various environmental summits under the UNFCCC process, efforts from different blocks of developing countries were made to accommodate their current stage of development deficit into the climate strategies. India signed the UNFCCC on 10 June 1992 and ratified it on 1 November 1993, in the hope that the developmental aspirations of developing countries like India would also be taken care of in the future summits and in any battle against environmental decline. In the early 1990s, India also signed many landmark conventions and agreements relating to environment. These include Montreal Protocol and the Convention on Biological Diversity (CBD). The latter was ratified by her in 1994. The UN Convention to Combat Desertification was also signed by her. She was also party to many other conventions like the Ramsar Convention and many more concerning wild flora and fauna and migratory species. The MoEF was the central body behind the implementation of these conventions.

The Earth summit was followed by COP-1 in Berlin in 1995 which was significant for India as it voiced her concern here against the developed world's demand. The US-led coalition including Japan, Canada, Australia and New Zealand (JUSCANZ) and EU countries led by Germany put forward the demand to introduce a new category in the environmental regime beyond the broad categorization of the developed/developing dichotomy. In order to ward off the Northern demand, India took a leadership position within the South by convening group of 72 'like-minded countries' from the G-77 (the so-called Green Group of developing countries led by her which lent their support to the Berlin Mandate and entered into a coalition with the EU) to cooperate with the Centre for Science and Environment and the Climate Action Network in drafting a 'Green Paper'. This paper called for negotiations on a climate protocol to be finalized at CoP 2 and for Annex 1 parties to adopt legally binding emission reduction targets within the context of this protocol.[9] In the negotiations for formulating Kyoto Protocol between 1995 and 1997, India worked closely with other G77 countries to exclude the concept of 'voluntary commitments' for developing countries which several developed countries had then tried to introduce.[10]

2.1.2. *India's Role in the Kyoto Negotiations and Beyond*

India by acceding to the Kyoto Protocol in 2002 had basically acknowledged the triumph of multilateralism to combat environmental change. Although along with other developing countries she did not face any mandatory emission cuts for the first commitment period of climate change, India and China refused to accept even voluntary commitments for themselves at CoP 4 in 1998 in Buenos Aires. But hosting COP-8, the eighth session of the Conference of Parties to the UNFCCC in 2002, India had shown her commitment to global climate regime. The Delhi Declaration as an outcome while reiterating the importance of implementing existing international commitment under the UNFCCC emphasized the links between climate change and sustainable development and prioritized adaptation concerns of developing countries and the need to promote technological advances through research and development. These moves were expected to substantially augment renewable energy use and to advance transfer of GHG mitigation options in all major economic sectors and substantiate market-oriented approaches.[11]

CDM, one of the flexible mechanisms of Kyoto Protocol, is of great significance for India. Even India is emerging as one of the chief beneficiaries of clean technology. India's support for the promotion and expansion of this mechanism was also evident when she accounted for over 20 per cent of total CDM projects registered worldwide by the UNFCCC by the end of 2010 with its private sector actively engaged in this process.

Although, in 1992, India published its first definitive report for the base year 1990 on an enlarged scale, it was only in 2004 that India submitted its initial National Communication to the UNFCCC keeping 1994 as the base year. It comprises inventory which acts comprehensively towards estimating India's emissions from all energy, agriculture and land use sources. Sectoral GHG emissions and vulnerability assessment are also part of the same.[12] At that time, India's annual emission were approximately 1228.54 million tonnes of GHGs, which was quite large but on per capita basis, her emission level was significantly low (28% of the global average and only 4% of the level of the United States).[13] Despite this fact, India in particular and the developing countries in general were targeted to take mandatory emissions cuts for their rising trend of emissions.

2.1.3. *India in the Post Kyoto Process*

The year 2005 was significant because Kyoto Protocol came into force this year and the process for building post-2012 climate regime began to take place. India also took crucial parts in those discussions under the aegis of the UNFCCC process, although simultaneously she engaged herself in several bilateral and multilateral fora on this issue even outside the UN like the US-led Asia Pacific Partnership. Doubts were raised at that juncture regarding India's commitment to global efforts to address environmental change under UNFCCC. From this time onwards, more pressure began to be applied by the North on emerging developing countries again, especially China and India to do more on climate change. As India's emission had grown by 97 per cent between 1990 and 2004, which was noted to be one of the highest rates of escalation in the world and China had in 2007 left the United States behind as the world's largest aggregate emitter of GHGs, a number of developed countries forcefully argued that no long-term solution to environmental change would be feasible without the participation of these fast emerging states especially like India and China.[14] Against such a background India had consistently been of the view that historical emissions were of great significance to establish equity under the UNFCCC. She asserted her climate posture in the post Kyoto process that began to take prominent shape in the Bali summit in 2007, endorsing that historical emissions and the CBDR are fundamental to Indian strategy which is non-negotiable.

Indian Government took a number of steps during that period domestically in order to show the rising significance that it accorded to tackling the issue of climate change. The National Environmental Policy (NEP) was formulated in 2006 which provided guidelines to various environment conservation and pollution control boards for environmental protection. India's response to environmental change was also outlined by it. NEP contains policies like India's adherence to CBDR, concerns for various climate-driven vulnerabilities and so on.[15] The Policy guided the governmental clearance policy for environmental and mining projects too. However, it was criticized for being a replica of United Nations Conference on Environment and Development (UNCED) documents as most of the principles were adapted from those enunciated in that document but how those have been internalized into prevailing policies, and introspection, unfortunately, was missing.[16]

In order to handle the external pressure against her for adopting more positive and transparent gesture to combat the problem, another significant effort had been made by the then PM Dr. Manmohan Singh, with the constitution of Prime Minister's Council on Climate Change (PMCCC) in 2007. The council held its first substantive meeting only in mid-2008 when it was presented with a draft of National Action Plan for Climate Change (NAPCC).[17]

The PMCCC was constituted just before Heiligendamm G-8 summit in Germany to which India was invited as an observer. At this G-8 summit, the Prime Minister again noted that India's GHG emissions in per capita terms were relatively low in contrast to the other major emitters. Such repeated attempts by the Indian government to talk in terms of per capita emission line of thinking of various global climate summits were often criticized as feet dragging approach, and there was tendency to tag India as 'climate denier'. However, the idea of voluntary cuts in emissions as propounded by the PM had been titled 'Singh Convergence Principle' that actually constrained India's emissions automatically. Thus, while India was announcing such unilateral and voluntary pledges to cut emissions, it actually gave an alert to the other major emitters to take a positive attitude in their climate postures beyond 2012.

The NAPCC had been released as a means for voluntary actions for climate change on 30 June 2008 with eight National Missions.[18] It covers both measures for mitigation and adaptation. The action plan tried to maintain India's developmental aspirations while reaping co-benefits from measures taken to alter environmental change. Along with the eight missions, the NAPCC also outlined 24 initiatives that would have substantial benefits in terms of addressing climate change, when integrated with the development plans of the ministries.[19] This plan of action basically had been designed to maintain a balance between the efforts to sustain India's growth while dealing with the climate change. The principle of equity in burden sharing was once again underscored here as India urged to ensure the equal entitlement to the global carbon space of each individual.

Before the formulation of the NAPCC, India also emphasized the need to incorporate this principle in the international fora like CoP13 at Bali in 2007 where she played a prominent role in the negotiation of the Bali Action Plan and finally agreed to take actions for reducing emissions that would be subject to verification, reporting and measurement but she refused to take any legally binding commitment in this direction.[20] She was firm on her plea that developed country would help financially for the actions taken to abate climate change. The year 2009 witnessed a 'flurry of diplomatic activity'. The G-8 summit in L'Aquila, Italy, Major Economies Forum (MEF), was significant among them that involved discussions on the topic. To India, they were equally significant as the then Prime Minister Manmohan Singh signed the MEF Leaders Declaration on Energy and Climate at a meeting held on the sideline of the G-8 summit in L'Aquila. But the G-8 summit triggered a controversy over India's signing of the declaration of the MEF. The Indian move to work together with the MEF countries for reducing global emissions by 2050 had been criticized as a compromise to India's previous non-negotiable stand. Critics argued it would hinder her development aspirations.

In the years preceding Copenhagen summit, there were efforts made by the developed world to abandon the Kyoto, a move that India thought would go against her interest. Although equity in burden sharing was the cornerstone of India's climate posture, in the run-up to the Copenhagen summit, there emerged series of domestic debates over the issue and re-framing of earlier Indian position that was not at all proactive. The then Environment Minister Jayram Ramesh put it more candidly when he tried to explain the need of being a proactive player in climate negotiations in an interview. He denied to make our domestic actions contingent upon international pressure and described the required approach as the per capita-plus approach.[21] So before Copenhagen the stage was already set for repositioning India's mandate as a leader, as a proactive player in order to shape the solution of the problem. There were flood of debates in the Parliament regarding India's changing role in the climate diplomacy as re-framed by Jayram Ramesh. Finally, a voluntary pledge was made just before Copenhagen Summit by India that it would reduce 20–25 per cent of 'emissions intensity' of its GDP by 2020 compared to its 2005 level through domestic mitigation actions.[22] This statement had added fuel to the ongoing debate across the country. In the face of such crisis, the Minister of Environment stated in the Lok-Sabha debate held on 3 December 2009, India's three non-negotiable stands in the climate negotiations. They were

1   Non-adherence to any legally binding emission reduction targets;
2   Non-acceptance of a peaking year for India under any circumstances;
3   Non-acceptance of external scrutiny to unsupported mitigation actions as those that are externally supported.[23]

The Left Parties feared that India might sell out her national interest since the government had adopted a stance that some flexibility was desirable in the negotiation. It was also said that announcing a voluntary pledge before the summit was not at all a good move. The Environment Minister strongly opposed this view by saying that flexibility only implied that India would take a positive gesture in the upcoming summit for paving the way towards forming an equitable burden-sharing atmosphere.[24] Amidst such chaos, Copenhagen Summit took place where India in coalition with a core group, called the BASIC, jointly resisted the mounting pressure of accepting stronger mitigation commitment by the US-led North. The developed countries were of the opinion to replace the Kyoto Protocol with a new agreement that would be applicable to all without making any differentiation.[25]

Copenhagen was expected to open up new avenues of opportunities for India to take the lead on the world stage. But she failed to achieve a comprehensive global deal. Although the Indian negotiators along with certain NGOs were happy that there was no peaking year limit set in the Copenhagen accord, there was a hue and cry on the MRV of national commitments made by developing countries, particularly India and China to reduce carbon or energy intensity of GDP.[26] The Accord required the countries like India to enlist mandatory mitigation actions that they would desire to commit by 31 January 2010, while stipulating the developed countries to submit under Appendix I to the Accord, 'individually or jointly quantified economy wide emission targets' by the same time. However, the mitigation targets once committed by them, and these actions would be subject to a certain degree of international scrutiny.[27] Against such a backdrop, what was required was that the government of India should draw a clear picture of the nature of mitigation actions it could commit to. It was accepted by India that her mitigation actions would be reported every 2 years for verification. Such a move was viewed as a compromise that was previously promised not to accept under any circumstances. Though the accord was not implemented finally, such changes in India's stand had been well recorded. Copenhagen summit witnessed a new hyphenation between India and China, both of which along with other G-77 countries wanted to expand their carbon space to maintain rapid emission intensive GDP in the name of defending their poor multitude. China refused any quantitative targets and India went along.[28]

Internationally after a much discussed and over hyped Copenhagen Summit with minimum tangible outcomes, the CoP 16 held in Cancun was deemed a success by the international community. It was seen by many including Indian analysts, a good foundation to move forward. Even before the Cancun conference, India worked hard and the government also prepared itself to promote her role as a 'deal maker' at the conference.[29] The Cancun Summit witnessed for the first time after Copenhagen deadlock, the positive dimension of such Indian effort.[30] The Copenhagen Accord and Cancun agreement had already attempted to blur the firewall, the Durban Platform for Enhanced Action, a measure taken in the Durban CoP of 2011, once again endorsed this idea. India strongly resisted this proposal as accepting a legally binding treaty would imply crossing the cabinet-mandated red line. In the Durban Summit, India preferred to maintain her defensive position in the climate architecture – that was to stick to the demand for the second commitment period of the Kyoto.[31] However, India ultimately assented to the Durban Platform without even the token inclusion of any of its core concerns such as equity for the first time.[32] Article 7 of the Durban Platform talked about equal efforts by all parties for mitigation to which India gave her nod.[33] Such urge for bringing a symmetry in commitment between the developed countries and emerging economies was the result of the EU and the US insistence.

At the Rio+20 meet of 2012, India again underscored the principle of CBDR and her priority in any climate meets to eradicate poverty. In the negotiation process at Rio, these criteria remained significant as the principle which was pushed aside at Copenhagen and Durban had found its way back in the outcome document. North–South rift once again came into prominence as the G-77 plus China in general and BASIC in specific raised their voice in unison at the summit. The Doha and Warsaw climate summit of 2012 and 2013, respectively, had proved that trend once again. In the Lima conference of 2014, the nations again did not reach consensus on finalizing an all-encompassing climate deal to be signed in Paris in 2015. India's strategy during that period was to prepare for the Paris summit where she would emphasize adaptation more than mitigation efforts.[34] India was consistently underscoring in that period the principle of CBDR and the operationalization of the Green Climate Fund (GCF).[35] For several years, funding crisis was plaguing the negotiations, so India showed in the conference her eagerness to break the deadlock.

2.1.4. *India in the Post-2020 Process*

In 2015, Paris conference, again India has shown its positive gesture towards multilateral climate cooperation. India formally adopted the Paris agreement in October 2016. Following the spirit of the agreement, India pledged to reduce 33 to 35 per cent of its GDP's energy intensity by 2030 from 2005 levels and also committed to achieve 40 per cent of its electricity generation from renewables by the same year. Regarding finance, it also proposed to mobilize domestic funds and new or additional funds from developed countries for the implementation of the mitigation and adaptation actions.[36] Since the Paris conference India is scoring high on the climate front, it has shown greater cooperation on climate action through its NDCs and projected its commitment to a sustainable, low-carbon future. India's positive role in the success of the Paris conference had been acknowledged by all, and PM Modi's initiative on the setting up of an International Solar Alliance for promoting solar power worldwide had also been acclaimed. In the Glasgow Summit of 2021, India has shown her positive climate posture for the post-2020 climate regime as well. India has committed to achieve net-zero emission target by 2070 at this summit. Though its emissions have added up to a miniscule 4 per cent of the global total only up to 2019, such ambitious pledges have marked a shift in her climate strategy which was often criticized as 'feet dragging'.

2.2. *The Security Council Involvement and India*

Regarding Security Council's role in dealing with the security dimension of climate change, like China and Russia, India also endorsed that the UNF-CCC was a more appropriate place for climate negotiations, because it includes all member states.[37] Even in the 2011 Security Council debate on the issue, India protested against UNSC involvement in the climate issue as she believed that it would infringe on the authority of the other bodies and compromise the rights of the Organization's wider membership.

Climate change and environmental disasters affect people, societies and ecosystems around the world, but the degree of vulnerability differs and as a result countries respond to the crisis induced by climate change differently.

Each nation's socio-economic conditions and its priority for development are also very specific which have determined largely the response of that particular country to the environmental threats. The previous sections of this chapter have discussed India's role in various climate summits and how the Constitution has endorsed the importance of environmental protection in some of its provisions, while the following sections will highlight how she responds to climate-induced human security issues.[38]

3. Indian Response to Resource Scarcity

Environmental degradation and climate change have serious bearing on natural resources in India. They have long-term effects on both economy and socio-political settings of the country. So along with playing a crucial role in the global battle to combat environmental threats, several domestic measures have been adopted to confront the problem of changing climate-driven paucity of resources. Meanwhile, in the landmark climate summits she often demanded measures related to resource scarcity related problems too.

3.1. Response to Lack of Food and Arable Land

Indian agricultural policies and some related programmes have tried to adapt to climate variability. There are various governmental schemes that the farmers can opt for like the Kisan Credit Card scheme, the National Agricultural Insurance Scheme and Scheme on seed crop insurance, to name a few. Some of them were launched to facilitate short-term credit to farmers, while some concentrated on providing coverage to the natural calamity-led losses. These are kinds of social protection schemes that were needed to encounter the natural calamity induced agricultural losses.[39]

Before the launching of NAPCC in 2008, the National Food Security Mission launched in 2007. It was a centrally sponsored scheme. The mission had the goal to sustainably produce the staples more in prime agricultural regions of the country in order to ensure food security as a whole. The environmental changes that pose additional threat to food security were also emphasized by the plan.[40] The National Mission for Sustainable agriculture (NMSA) as a part of NAPCC basically aimed at developing climate resilient crops, expanding mechanism for weather insurance and weather-specific agriculture. All of these would support adaptation to climate change. Dryland agriculture, use of biotechnology in agriculture, managing risk and information access were also focused to be discussed by the Mission.[41]

The mission document on NMSA tried to implement a programme of action (POA). It covered both adaptation and mitigation measures. It also emphasized the need to use traditional knowledge base and agricultural heritage of the country for establishing sustainability in agriculture.[42] The NMSA also focused on the use of advanced biotechnology, including genetically modified seeds; introduction of a C-4 photosynthesis pathway into

C-3 crops like rice, wheat and groundnut; and mining of genes involved in high nitrogen and water use efficiency. So various steps were made under the NAPCC for the sustainable and climate resilient agriculture. Some measures were also taken to guarantee social protection for the farmers in the face of climate-induced crop failure and resultant food scarcity.

But the NMSA pursue the path to reinvest in energy and input intensive agricultural production and is dictated by the demand for agricultural production enhancement for national food security which ultimately pushes agriculture against environment. Production policies and technologies also contribute significantly to the economic and ecological distress in rural agrarian India. The technologies used and interventions for assuring food security are often unaware of the location and interaction of farming with specific environmental and social context. The new ways of food production may engender unemployment, environmental disruption and food crises for the rural poor in the long run.[43] The NMSA though tried to minimize the vulnerability of agricultural sector, it was not aware of the fact that in a country like India, agriculture could not be dealt with in isolation from economic issues. Seldom there were practices of dealing with environmental change and economic globalization in conjunction.

With economic globalization, farming is now a corporate venture and the corporations dispose of the rights of the nature. The seeds supply in India was taken over by the MNCs since Green Revolution. One instance was the replacement of locally grown mustard oil despite its abundance with imported soybean oil. In 1998, protests erupted across India against the 'Soy Imperialism' when the plans for importing 1 million tonnes of US soybeans to be used as oil seeds had been announced. Ironically at that time mustard seeds can be processed locally, making it available to the poor at low cost.[44] Apart from the negative economic effects, it has been said by environmentalists like Vandana Shiva that GM crops are basically unsustainable and often affect the species which are beneficial for land and crops and they destroy soil as well, thereby affecting the environment.[45] There is no guarantee whether these processes would yield better results than reliance on traditional practices of crop improvement, including the use of sturdier, drought-resistant seeds and low-energy low-input farming. But the NMSA was not vocal on this account. It did not recommend a decisive and rapid transition to low-energy, low-input agriculture based on indigenous grains and seeds, sound watershed management practices, prudent methods of moisture conservation and prevention of soil erosion and high water runoff, all the way to zero-tillage organic farming. Instead, it recommended alternative crop rotation cycles for different regions, but without reducing excessive dependence on wheat and rice under a predominantly BAU scenario, and without major changes in existing growth strategies and priorities.[46]

Subsidies and pricing mechanism in agriculture are also causing detriments to the environment. Subsidies for fertilizers have augmented the use of urea, leading to nitrogen–phosphorus–potassium imbalances in the soil. The subsidy regime is responsible for that as farmers using urea at a comparatively lower price leading to their overuse.[47] Along with this, the production policies and technologies are not farmer friendly causing both economic and ecological distress in rural India. Most of the farming communities are incapable of using highly expensive modern agricultural inputs like fertilizers, chemicals water seeds and so on. Finally, the mission document is also silent about the coastal livelihoods. Climate change induced sea level rise and the resultant increasing salinity of the coastal lands do not figure in the NAPCC.[48]

Along with the governmental activities, there exists a spurt in NGO activities in establishing environmental awareness. In fact, various NGOs are working in this direction since long. Agriculture and natural resource management are their area of interest. The M.S. Swaminathan Research Foundation which is a Chennai-based non-profit NGO took up projects related to climate change adaptation measures in Indian agriculture. They carried out fieldwork in Andhra Pradesh and Rajasthan and suggested transformation in the existing water management system, using weather data from farmer-controlled agro-meteorological stations, sharing knowledge through smart farmers' network, etc.[49] Furthermore, in order to keep food production in the hands of farmers and away from corporate control, some NGOs have campaigned for ecological farming which helps cope with climate change. The Greenpeace India has campaigned for it. Eco-farming helps in conserving soil and water and helps in accelerating diversity in agriculture and also in the biosphere.[50]

The pandemic, which started spreading its tentacles in India in March 2020, affected the agricultural output tremendously. In such a situation, three new farm laws were announced by the Central Government in September 2020, as part of economic recovery. The new farm laws are the Farmers' Produce Trade and Commerce (Promotion and Facilitation) Act, 2020, Farmers' (Empowerment and Protection) Agreement of Price Assurance and Farm Services Act, 2020, and the Essential Commodities (Amendment) Act, 2020. The three laws aimed at reforming the agricultural sector, but the farmers were extremely disappointed with these legislations as they believe they did not guarantee the minimum support price (MSP) while leaving them at the whims of corporate giants. As a result, the chances of government agencies to procure farm outputs get diminished. The impact of climate change is integrally related with MSP as changes in climate and the resultant crop failure affect the earnings of the farmers. Famers became agitated throughout the country in the demand of repealing these laws, and in November 2021, these three farm laws were repealed.[51]

Food scarcity and drops in production are also linked to desertification and land degradation. India has taken several steps to combat this threat too. As desertification is a problem which is very much associated with resource scarcity, many efforts to check the latter automatically address the problem of land degradation and desertification like National water policy and National environmental policy. India signed the UNCCD on 14 October 1994 and ratified it on 17 December 1996. She also prepared a National Action Programme (NAP) to combat desertification in 2001. Given the cross-cutting dimension of desertification, the NAP recognizes the multi-sectoral nature of the task. As along with climate change poverty and unsustainable use of land are greatly inducing the problem, NAP's objective was to follow an approach which would be community-based. So self-governance and empowering the local community was one of its strategies.[52]

Despite such steps taken by the Indian Government, it is true that there is no way we can meet food security unless the marginalized sections get opportunity to come out of the poverty trap on the one hand and the proper implementation of sustainable pro-environmental agricultural practices take place. There is a plethora of environmental plans and programmes, but their success is programme-specific. The integrated approach to tackle the problems is missing. For instance, if we require to counter decline in agriculture due to climate change, other factors should be taken into considerations which are related to drops in agricultural output like water scarcity, land degradation as well as the policy gaps to address this nexus.

3.2. *Response to Water Scarcity*

There exist several constitutional provisions, laws at both the centre and state levels as well as court dictums that have created the structure of water laws in India. It is a state subject, so the governments of the states have the right to control its use.[53] Article 21 of the constitution implicitly guaranteed the right of the people to receive pollution free water. So it is the overarching responsibility of the Indian government which was endorsed by the Supreme Court also to create a condition so that each one can get access to clean potable water.[54] However, there are lacunae in implementing the unequivocal right to water in India.

There are several legal documents, though they are not directly related to the measures combating climate change induced water scarcity, they are useful for managing this dwindling resource. The most significant in this regard was the National Water Policy (NWP) that had been adopted in 1987 and revised in 2002 by the Union Ministry of Water Resources. It required also the formulation of respective state water policy that would work under NWP. The scarcity of water as a natural resource had been recognized by the NWP. It underscored the importance of sustainable and regulated groundwater exploitation as well as the management of irrigation systems and more importantly emphasized the need to shift focus from water project to resource policy issues.[55] Efficient management of water also found expression in the 11th Five-Year Plan (2007–2012). The launching of the Jal Abhiyan Programme in 2005 was part of it. It tried to enhance awareness regarding surface and groundwater management, the recharge of groundwater, etc. Almost 20000 villages were covered with the development of many water-harvesting structures and revamping of canal system under this plan.[56]

However, such policy measures are not enough. There should be methods to counter the problems associated with climate change and water scarcity nexus. To address the problems pertain to groundwater overuse and its debilitating quality, a process-based approach which would address the impacts of climate change on water together with the rising demand for this depleting resource was actually needed.[57] Due to the open accessibility of groundwater, extraction of it is mostly unsustainable and its resultant effect is found in the rising decrease in water table in various states of India. So to ensure its sustainability, the Indian Government has tried to find ways time to time. For regulating and controlling development of groundwater, several orders had been passed in 1996 by the Supreme Court. They recommended the setting up of Central Ground Water Authority (CGWA) under the Environment (Protection) Act, 1986 which was delegated to control the groundwater depletion.[58]

As there is growing concerns regarding climate change in the popular discourse in India, the awareness to conserve water has also started gaining momentum. The National Water Mission (NWM) was a fruit of that growing consciousness of water scarcity. It was also a component of NAPCC. The main objective of the National Water Mission was the conservation of the resource, reducing its wastage, distribution of water equitably through integrated management of the resource and so on.[59] However, the mission is not beyond criticisms. No doubt it encompasses a wide variety of ways to address the crucial relation between climate change and water like setting targets for achieving water efficiency, promotion of traditional system of water conservation, recycling urban water and so on, but it does not set any plans, benchmarks or programme framework for achieving this. It tried to assess the climate change impact but not devised strategies to combat the climate challenges. Glacial retreat, increasing frequency of floods, decreasing flows of river water, sea water intrusion are inevitable effects of climate change, but the Mission is quite silent on how to deal with these problems. Critics have pointed out that the Mission has failed to analyse the negative impacts of India's big water projects like big dams and canals. The document's continued advocacy of large dams as the best possible way to meet the climate change induced rising water demand is also a matter of criticism.[60] The Mission is also criticized on the ground that it is devoid of community involvement. Some sections considered the NWM as non-transparent and non-participatory.

But the awareness about environmental change induced water scarcity is growing as the time passes. The 2012 draft NWP again underscored the grim reality of both natural and human-induced water scarcity in India. It reiterated that India has over 18 per cent of the world's population, but barely 4 per cent of the renewable water resources. Per-capita water availability has steadily declined and is further expected to drop in the future. Further, it also noted that a huge share of discharged waste water into rivers was not really 'treated' at all, even as there was rising unsustainable dependence on groundwater for both domestic and irrigation purposes. Additionally, despite rainfall concentrated in only a few weeks in a year, we lack sufficient storage infrastructure; its maintenance is also routinely neglected. Hence, there is vital need for recharge of aquifers and rainwater harvesting.[61] After a decade, given the fact that availability of usable water is coming under further strain, this new NWP recognized the variability of water resources due to climate change. Among the adaptation strategies the plan suggested to escalate storage of water in its various forms. As in India agricultural sector uses a large share of country's water resources, compatible strategies are required to be adopted in the yielding pattern. By involving water users while sensitizing them to the process and by building capacity, this process can become successful.[62] Although the industry association who lobbied for the draft have welcomed it, from several sections it was criticized for being a slight improvement from the previous NWP of 2002. Although implicitly it acknowledged that water requirement for the survival of human beings and ecosystem precedes the need for water use for economic good but as it was not explicitly enshrined in the draft, conflict over different users was very likely.

To update the current NWP which was last drafted in 2012, an 11-member committee was formed in 2019 by the Union Jal Shakti Ministry. This Mihir Shah led committee found that still water-intensive crops like rice, wheat and sugarcanes are grown in relatively water short regions of India which contributed more to the water scarcity of the country. The committee also suggested crop diversification to get rid of such problems along with building Independent Water Resources Regulatory Authorities (IWRRAs) in all states and UTs.[63] The Government's Procurement Policy for wheat and rice had also been blamed by the Draft National Water Policy of 2020 as it aggravates the water crisis in the country. Besides, in order to prevent the farmers to grow rice and wheat, the National Water Policy 2020 has suggested to fix the Irrigation Water Fee on Volumetric basis which means farmers have to pay for the amount of water they use instead of paying a fixed amount.

The pollution of river water due to dumping of wastes is another matter of great concern in India. The Ganges, the lifeline of India, is severely being polluted by towns and industries. The National Environment Engineering Research Institute was commissioned by the Government of India to study the sources and causes for water pollution along the entire course of the Ganges from the Himalayas to the Bay of Bengal. The report of this study that identified several causes of pollution in the river was placed before the Supreme Court. Judge Kuldeep Singh ordered the GoI to create a special plan of action that would ensure that the Ganges was cleaned up within a specified time. This judgement was a landmark one and served as the basis for adoption of the Ganga Action Plan (GAP).[64] Phase 1 of this plan was launched in 1985 for improving the water quality of Ganga, and it was completed in 2000. The Phase II of the programme included its tributaries.[65] The Narendra Modi Government also emphasized the need to rejuvenate the Ganga. The term Ganga Rejuvenation had also been incorporated within the name of the ministry of water resources. Demands are raised from various quarters to the government to intervene and ensure that the objective of reviving the dying is not compromised by various sectoral plans, programmes and projects.[66] Rivers become more significant when they are shared between two states of the same country or between two countries. There are numerous treaties to resolve the shared water disputes between India and her neighbours. However, they do not adequately respond to the environmental and political challenges posed by water conflicts. This is a critical situation for India as regionally she is encountering resource scarcity generated political battle with her neighbours while at the same time, globally, she, along with them, has to raise her voice against the centrality of inequity in the sharing of climate change burden. Countries therefore must go beyond the narrow confines of self-interested behaviour as climate change respects no border. Regional political turmoil should not impede the efforts to reach a consensual deal for addressing the nexus of environmental change and shared water resources related problems.

3.3. Response to Deforestation

Deforestation is a glaring problem in India. There are number of laws, orders and institutions that directly concern the sustainable management of forestlands in India and try to protect it from degradation. The first Indian Forest Act was codified in 1865. It witnessed several amendments and ultimately led to development of Indian Forest Act (IFA) 1927. It prohibited the use of forest land for non-forest purposes.[67] Other legislations related with IFA 1927, like the Forest Conservation Act, Wildlife Protection Act, the Environment Protection Act and the Biological Diversity Act, are also critical in framing the legal structure for the protection of forest and its parts like wildlife and the biodiversity.[68] Among others, the National Forest policy of 1988 is equally important. It entails the participatory approach to forest management. Along with maintaining the ecological balance and sustainability of forests, it entails the relation between forest and the forest-dependent communities, especially the tribal people and their customary rights to the forest. It talked about their livelihood security as well.[69]

Large afforestation and reforestation programmes has also been promulgated in India since 1980. The National Afforestation programme (NAfP) was launched in 2002. During the period 1980–2005, the cumulative afforested area amounts to nearly 34 million hectares.[70] To address the fundamental problems of the forest sectors, a comprehensive strategy was required. This was implemented by a National Forestry Action Programme of 1999 that would work in line with NFP.[71] Along with these efforts, some institutional arrangements were also made. Technical capability to assess the forest cover had also been developed over the time. The Forest Survey of India (FSI) is doing such jobs. The forest carbon stocks accounting for the country had been done by this institution.[72]

In order to recognize the close relationship between forest and the Indigenous communities living in and around the forest, the Joint Forest Management has been introduced in 1990. It initiated the process of involving local communities in the protection and management of the forests.[73] This JFM introduced a system of decentralization in decision-making where both the forest department and village community work in unison and jointly manage the forestlands.[74] The Forest Rights Act (FRA) was another landmark legislation which had tried to strengthen the basis of JFM. It was enacted in 2006.[75] Later JFM was transformed to JFM+. The livelihood concerns of the forest-dependent communities are incorporated through this.[76] JFM+ has been structured in such a way so that traditional knowledge base can be blended with the scientific methods while managing the forest. By providing livelihood security to the forest-based communities, JFM+ actually paved the way for the emergence REDD+ platform in India.[77] But JFM has been criticized because proper decentralization does not take place in reality. The roles of village communities are still left to the peripheries.[78] The transfer of the rights to manage forests to the Gram Sabha under FRA actually downsized the JFM hugely.[79] As JFM was introduced to incentivize the local community to protect the forest nationally, at the international level, REDD emerged.[80] However, REDD+ has been advocated by India later to make REDD more comprehensive. It demanded rewards for afforestation projects also.[81] REDD+ is basically strengthening the JFM. It was expected that the local communities would receive the incentives generated from REDD+ that would inspire them to work towards the sustainable protection of the forest lands.[82]

However, REDD plus is not beyond criticism. It is argued that often the afforestation programmes take place on cultivated lands, village commons and community pasture lands leading to evictions of local people and to a wave of land grabbing for plantation programmes which usually replace grasslands, scrub jungles and other habitats with monoculture.[83] Often the monocultures affect the water cycle as was the case in Deccan Plateau where streams and tanks dried up due to the spread of eucalyptus monocultures. Such man-made ecological stress gave birth to unrest as well. In 1983, farmers in the state of Karnataka marched to the forest nursery in order to uproot millions of eucalyptus seedlings and planted tamarind and mango seeds in their place.[84] In Purulia, Bankura, Midnapore, Singhbhum and Palamou, the eucalyptus monoculture also affected the water resource required for drinking and irrigation.[85] Finally, the carbon trading model inherent in REDD mechanism may actually create larger financial incentives for wholesale takeovers of forests.[86] Given such inadequacy and loopholes in the post Cancun period development of forest had been connected to the Green India Mission.

India's concern for the protection of forest resources were once again expressed through the launching of Green India Mission. Its aim was to augment the forest cover and the density of the country. It also emphasized the need to enhance the carbon sinks and initiated afforestation projects also.[87] Not only dense forestlands, afforestation in scrubland, mangroves, cold deserts and areas for shifting cultivation as well as deserted mining areas are also outlined in the Green India Mission.[88] It might provide opportunity to finance the REDD+ projects as well.[89]

The Green India Mission was very significant, and it recommended amendments to the Indian Forest Act and the Panchayati Raj Acts in order to empower the JFM committees as the forest officers who would act as the legal entities of Gram Sabha. Actually, the mission tried to strengthen JFM. But mere mention of promises failed to act in reality. It did not emphasize forest conservation. Through JFM initiatives, efforts were made to prevent deforestation or promoting forest conservation earlier. But there was failure to reach these goals. The mission document was also silent about those failures.[90] The mission is also criticized as it does not include two of the most vulnerable forest regions of India- the Himalayan Foothills and slopes and the Western Ghats. It contains elaborate details of the organizational and institutional arrangements through which the programme would be implemented. But it is remarkably inefficient on two central issues. First, it fails to reform the Forest service bureaucracy which remains hostile to people's participation in forestry programmes and to implementing the forest rights; Second, it does not give priority to the non-implementation of Panchayati Raj Extension to Scheduled Areas (PESA) Act which can give shape to the noble thought of community involvement.[91]

Management and protection of forest have entered the realm of national priority today. She has therefore taken several steps to protect the forest resources as a part of its effort to combat environmental change. But forest and wildlife are inseparable. So protection of wildlife is inextricably linked with forest conservation. Several legal instruments help conserve the wildlife in India also. This includes provisions related to building sanctuaries and national parks for the preservation of wildlife as enshrined in the National Forest Policy 1952 and Wild Life (Protection) Act 1972, and many more. There are certain projects that also protect the wild habitat of the country. Among them Project Tiger (1973), Project Elephant (1992) and the Biological Diversity Act (2002) were important. The Wildlife (Protection) Act 1972 also presented a platform to conserve forestry with the constitution of a series of national parks and wildlife sanctuaries.[92] Apart from the protection of wildlife, need was felt that measures should be taken to save the diverse biological heritage of India which is vulnerable to environmental change and human activities. The next section has shed light on it.[93]

3.4. Response to Biodiversity Loss

Loss of biodiversity is alarming for the ecological balance of any country. Significant deterioration in the biological heritage of India therefore requires intense concerns. An effective instrument towards this direction was International Convention on Biological Diversity (CBD). It was signed by India on 5 June 1992, and ratified on 18 February 1994. Table 4.1 shows how India has responded to the crisis over the years.

These efforts have revealed India's awareness of the danger associated with loss of biodiversity. In line with the commitment in Nagoya, like other countries she updated her national biodiversity strategies and plans of actions to reach the Aichi Targets.[94] Her ratification of the Nagoya Protocol has also expressed the country's urge to take meaningful leadership in the effort to prevent bio-piracy.[95] There is no doubt that changing climate has contributed largely to the loss of biodiversity but adopting the industrial agricultural model and development of livestock have also contributed to the destruction of the bioresources. In India, such human-induced biodiversity loss is very rampant. The corporate giants have started monopolizing the supply of seeds and genetic resources, thereby depriving the real custodians of genetic wealth as well as prioritizing the commercialization of seeds over conserving diversity. Due to the spread of the regime of monocultures,

Table 4.1 Steps Taken by Government of India to Conserve Biodiversity

Years	Steps Taken by Government of India to Conserve Biodiversity
1999	• National Policy and Macro-level Action Strategy on Biodiversity was prepared by the government
2000	• National Biodiversity Strategy and Action Plan (NBSAP) was implemented by the MoEF. It was externally funded
2002	• Biological Diversity Act had been enacted. Its aim was to conserve the biological resources
2003	• National Biodiversity Authority had been established by the Central Government
2006	• The National Biodiversity Action Plan (NBAP) was prepared
2008	• The Union Cabinet approved the revised NBAP on 6 November 2008
2010	• The Nagoya Protocol on access and benefit sharing (ABS) had been adopted in 2010
2011	• Nagoya Protocol had been signed by India on 11 May 2011
2012	• India's ratification of Nagoya Protocol was approved by the Union Cabinet on 9 October 2012 • The 11th meeting of the Conference of the Parties (COP 11) to the Convention on Biological Diversity (CBD) had been hosted in Hyderabad by India

Source: "India's Fifth National Report on Convention on Biological Diversity" Ministry of Environment and Forest, Government of India, 2014. Available at http://www.cbd.int/doc/world/in/in-nr-05-en.pdf Accessed on July 26, 2014.

we have lost not only the diversity of crops but along with this, the basic capital of soil and water is at severe risk due to excessive use of pesticides and fertilizers. In the face of such a crisis chartering, a new path towards conserving biodiversity is, therefore, a bare necessity in India.

3.5. Response to Problems Pertaining to Energy Resources

India is facing multi-pronged challenges regarding energy sector. So it is trying to formulate policies that may promote the energy efficiency along with policies stressing the use of renewable energy technologies for tackling the problem of inadequacy in supply, inequity in distribution and environmental harms associated with fossil fuel combustion.

Building energy efficiency contributes towards emission savings. The Supreme Court order for the use of compressed natural gas in public buses, taxis and auto rickshaws, in order to improve the air quality is another positive move to this direction. Coal substitution is another advancement as it has broken the monopoly of coal use which was earlier the mainstay of commercial energy. Recently, many sectors are switching to use natural gas instead of coal like the power sector, fertilizer plants and railways. The substitution of non-commercial energy like fuel wood, crop residues and animal dung, with commercial energy in cooking, lighting and in small-scale industries leads to increase in overall efficiency, but often consumers are shifting from old biomass to fossil fuels so the inevitable result is the increase in net GHG emissions.[96]

The legal basis of such efforts came in 2001 when the country has enacted the Energy Conservation Act envisaging energy efficiency standards for nine energy-intensive industries. It came into operational in 2002. The Bureau of Energy Efficiency (BEE) was also set up for developing strategies to fulfil the primary objective of reducing energy intensity of the Indian Economy. A standard programme for end-use equipment had been initiated by the Ministry of Power, through BEE.[97] The BEE has also initiated energy efficient programmes in various central government buildings and establishments.[98] The Planning Commission's Integrated Energy Policy (IEP) also provides an overarching framework to consider all aspects of energy use. It envisages conservation and efficiency of energy particularly by measures on the demand side.[99] The direction of India's Eleventh Five-Year Plan (2007–2012) was influenced by the IEP. The Plan highlighted the lack of sufficient energy supply and the coal dominance in the energy mix. There existed the dire need for the expansion of energy resources by increasing energy efficiency, exploring new sources, switching to renewables and applying contemporary R&D.[100]

The urge for improving energy efficiency as a strategy for mitigating climate change is also expressed in the NAPCC. The latter acknowledged the close nexus between Climate Change and Energy Security which is evident in its eight national missions that includes with other missions, missions on

Solar Energy, Enhancing Energy Efficiency and creating a Sustainable Urban Habitat. It underlined also the need to phase out coal. Before the Paris Conference India's voluntary pledge to decline the energy intensity of it, GDP was also commendable. India's positive attitude to enhance energy efficiency was quite pronounced within its INDCs also. Building 175 GW of renewable energy by the year 2030 had also been set as target. The success of the National Solar Mission is a positive step towards this. It has also been committed by her that no new thermal plants which are not of the most efficient ultra-supercritical category would be built.[101]

Carbon capture and storage (CCS) is one of the options to stop emissions from coal-fired power stations. The Indian CO_2 Sequestration Applied Research (ICOSAR) network had been launched in 2007 by the Department of Science and Technology. It facilitated dialogue with stakeholders for developing policy frameworks. The government has also started recognizing the benefits of CCS. A National Mission on Clean Coal Technology had been launched under NAPCC in February 2012.[102] However, there exist some roadblocks in reaping the full benefits of the CCS. The main constraint is financial,[103] transport and storage. In India, four facilities were found. They have been operating for decades in industries such as chemicals and fertilizers, where carbon dioxide is recovered from flue gas and used to manufacture byproducts.[104] A notable example of CCS technology was found at a plant in the industrial part of Tuticorin. It provided the first unsubsidized industrial scale example of CCS in India where CO_2 has been captured from its own coal-powered boiler and was used to make baking soda.[105] In the Glasgow CoP of 2021, phasing out of coal had been promoted by many countries. Though India was pressing to replace the phase out to phase down from the final draft of the agreement, its bold and ambitious target for non-fossil fuel power capacity announced at the summit has been praised by all.

India's renewable energy programme had started since the early 1970s. India exclusively have the Ministry of Non-Conventional Energy Sources (MNES) now known as Ministry of New and Renewable Energy (MNRE).[106] The 2003 Electricity act is another milestone in proclaiming India's commitment to renewable energy.[107] The renewable energy sector in India is dominated by wind and solar energy. The Jawaharlal Nehru National Solar mission with a target of 20 GW solar power generation by 2022 under which a majority of solar projects are located in Rajasthan had already been launched.[108] The NAPCC is also a positive move in promoting renewable energy, and it suggests increasing the renewable energy contribution to 15 per cent of electricity generation by 2020. The National Solar Mission is significant in this regard.[109] Hydropower is also an important source of clean and renewable energy. But hydro-power's share in India's energy mix is limited, and it experienced massive decline in the early part of 21st century in comparison to its level during 1960s.[110] India also encourages the use of biofuels and has formulated already its biofuel policy in 2009.

A comprehensive programme on biofuels for surface transportation had been initiated by the MNES since 2002. The aim was to build the technology for transforming vegetable oils and use then in the automotive sector.[111]

Against such a situation, another viable option for generating energy comes into prominence – the nuclear power. The Energy Information Agency observes that the bulk of the expansion in the nuclear sector is going to come from Russia, China and India, accounting for nearly two-thirds of net increment in nuclear power capacity between 2005 and 2030.[112] India's future energy generation is dependent on thorium-based nuclear power. India produced an experimental fast breeder on the thorium route (Kamini), and it works at Rameshwaram. Although the nuclear power is the only near-zero carbon electricity source aside from hydroelectric power, it has some shortfalls and negative impacts. Nuclear power plants in India have also met with severe protests from the local people of the project area given its radiological detriments. Protests against the commissioning of Kudankulam Nuclear Power Plant is one of such kinds.[113] People's fear of nuclear disaster was ruled out by the government, and they justified the project on the ground that nuclear energy would actually help enhancing the welfare of the people and would be beneficial for economic growth as it is an important element in India's energy mix, capable of replacing fossil fuels.

During the Glasgow summit of 2021, in order to achieve net-zero emissions by 2070, Prime Minister Modi has announced a very ambitious plan for renewable energy developments. It is a fivefold strategy, namely, 'Panchamitra'. It includes the following targets:

- Non-fossil energy capacity will be escalated to 500 gigawatt (GW) by 2030
- Renewables will meet 50 per cent of India's energy requirements by 2030
- One billion tonnes of total projected carbon emissions will be reduced from now onwards till 2030
- 45 per cent of carbon intensity will be reduced by 2030

All of these would help India to achieve the target of net zero by 2070.[114]

There are several weaknesses in Indian efforts to clean dirty energy usages. But the situation is not always that grim. The Narendra Modi government in India has played a critical role in broadening the range of actions in order to combat environmental change and its impact among which efforts to use more clean energy resources are significant. The Prime Minister has streamlined energy decision-making and also environmental decision-making. A new low-carbon approach to development has been suggested by the PM which underscores India's leadership role in the environmental parleys without compromising her national interest.

Resource scarcity is one of the problems or the risks that is exacerbated due to climate change but extreme weather events and rising sea levels are also affecting India's socio-economic fabric. The result is growing insecurity among the masses that has emanated from the fear of losing the natural rhythm of life and livelihoods. The former can be mitigated as discussed earlier through various measures that would reduce greenhouse gas emissions and increase sinks, thereby minimizing environmental change. But natural disasters and slow-onset process requires some more actions like behavioural changes, beginning with individual actions and ranging to collective coastal management policy, such as upgraded defences and warning systems and land management approaches.[115] India has therefore taken some of such steps to minimize these threats and also tried to improve the early warning system and relief actions so that the intensity and magnitude of natural disasters can be controlled to some extent.

4. Indian Response to Extreme Climate Conditions

India is suffering from both silent and loud environmental disasters. The adaptation measures that are taken by the country for them consist of three components like efforts to protect, accommodate and retreat.

4.1. Response to Sea Level Rise

India has a vast coastline. Therefore, it requires effective policy responses against the SLR. The Coastal Regulation Zone (CRZ) Notification had been issued by the Ministry of Environment and Forests in 1991 under Environment (Protection) Act, 1986.[116] The notification imposed restrictions on the setting up and expansion of industries or processing plants in the said CRZ. The CRZ was classified into four zones.[117] Among them, the CRZ-I that includes mangroves is the most vulnerable. The construction activity in this zone is also very limited. Projects relating to Department of Atomic Energy, laying pipelines, installing transmission lines by the India Meteorological Department are some of the exceptions that can be done in this region according to the CRZ Notification. But this notification was amended many times over the years. The Coastal Management Zone (CMZ) which was a new legislative framework had been brought by the environment ministry in 2006. CMZ had also been criticized by the environmentalists for having pro-industry-builders bias.[118] Before CMZ, the MoEF set up a committee under the Chairmanship of M.S. Swaminathan. The objective of that committee was to review the framework of coastal regulation. A more scientific way was required, the committee found. Accordingly, a proposal was made by the MoEF. It recommended a setback line based on the fragility of the coast to the sea level rise and shoreline changes. Such proposal was a significant departure from the previous line of proposals and corresponding notifications.[119]

Later, the CRZ Notification and Island Protection Zone (IPZ) Notification had been issued by the ministry in 2011. Its focus was to promote livelihood security for the local inhabitants.[120] The Island Development Authority cell had also been set up by the government. Its goal was to prepare Development Reports for both Andaman & Nicobar Islands and Lakshadweep. In coalition with the World Bank assisted Integrated Coastal Management (ICZM) Project, mapping of eco-sensitive areas along the mainland of the country including that for the island States had been initiated by the ministry also.[121] In India, this ICZMP started in 2010 as a follow-up to the recommendations of the Swaminathan Committee, in order to promote security in the coastal community and protect the ecosystems in a sustainable manner. With a budget of $285.67 million, the project plans to assist Indian government in developing an ICZM approach.[122] As accelerated sea level rise has affected the lives and livelihoods of coastal regions severely, such coastal management programmes are very significant as adaptation measures. Mention must be made here that in India, in order to protect coastal ecology and livelihoods, the Coastal Zone Regulation (CRZ) Notification, 2011 is very significant. Its aim is to manage the coasts and to protect the ecology thereof. The CRZ notification also helps in protesting against unscientific and indiscriminate development works at the coastal zones of India. Despite the existence of such regulations, Indian coastal zones are still fragile to climate change.[123] Cyclone Foni, Amphan and Yash have proved that again.

4.2. Response to Increased Intensity of Natural Disasters

Disaster management occupies a significant place in India's climate policy architecture. Since independence, a well-established framework for responding to disasters with strong relief measures and rehabilitation process has been put in place in the country. The institutional structures involved in the process include the Ministry of Home Affairs, that is, the nodal ministry for matters pertaining to disaster management, the Central Relief Commissioner (CRC) of the nodal ministry acting as the nodal officer for organizing relief operation and the National Crisis Management Committee (NCMC). India is a federal country. In this architecture, the responsibility associated with rescue, relief and rehabilitation lies in the jurisdiction of state government. The Union Government assists generally wherever it is required.[124]

In 2005, the Disaster Management Act was passed by the Indian Government. The National Disaster Management Authority (NDMA) has been established following the mandate of this act. A State Disaster Management Authority has also at the state level which is followed by Disaster Management Author at the district level as well.[125] Though there is a decentralized structure for disaster management, the NDMA is the apex body that guidelines, issues approvals, coordinates, and implements plans and methods required to build disaster preparedness at all levels.[126] The ambit of this act has been expanded when it was invoked during the COVID-19 pandemic in 2020 to impose lockdown. The central government has described this pandemic as a notified disaster, and for the first time in India a biological disaster has been controlled through legal and constitutional institutions.

A better-prepared community is a pre-requisite for disaster management. The Disaster Management Act also recognized the significance of the same. In India, the GoI – UNDP Disaster Risk Management Programme had first introduced preparedness at the community level in 2000. The GOI-UNDP Disaster Risk Management (DRM) programme (2002–2009 and 2009–2012) is instance of such effort. There are also non-governmental efforts towards this direction.[127]

The Disaster Management Act of 2005 also unique in its efforts to replace relief oriented response to disasters. Rather it emphasized on proactive measures coupled with early preparedness and mitigation-based approaches. The National Policy on Disaster Management (NPDM) had been prepared in tune with disaster management act for this reason.[128] Glacial retreat, its out bursts and flash floods are serious matters of concern for India today. To address the associated issues with this, under the aegis of NAPCC, the National Mission for Sustaining Himalayan Ecosystem was launched.[129] The urgency of creating national capacities to respond to the changes in a sustainable manner had been identified by this mission.[130]

Despite these efforts, unscrupulous and unchecked human actions have contributed to the increasing frequency of disasters in India. Changing climate has exacerbated the problem. In the governmental response to disasters, a better-prepared community is the most effective means for reducing the risks of the disasters. Though there exist community-based disaster management as envisaged by the NDMA, it has been overlooked and the government is often ignorant about the fact that the Indigenous people have greater knowledge of the environment and are capable of sensing the signals from natural processes occurring around them. This is evident during the 2004 Tsunami when Jarawas could sense the impending tsunami from nature's signal and made their way to higher land. So India should cultivate their knowledge while preparing her agenda for community-based disaster management.

5. Indian Response to Environmental Change Induced Displacements

The poor and weak sections of the society are the hardest hit of the climate change induced various threats as mentioned earlier. Drops in agricultural production, water scarcity, deforestation, extreme weather events and sea level rise all have affected their lives and livelihoods and as a coping strategy often they have opted temporary or permanent migration. Migration is a consequence of climate-driven threats, but as mentioned earlier the international law does not recognize these climate migrants in legal terms. In India, it is also difficult to distinguish the climate migrants from other types of refugees as socio-economic factors as well as political pushes can be simultaneously present as catalyst factors behind the migration. Yet the climate migrants are equally eligible to exercise the human rights as proclaimed by the Universal Declaration of Human Rights. But the government responses to this kind of climate migration should be more pronounced. Relief operations are fruitful, but these are only post-disaster activities. What is more important is to devise strategies to improve the early warning system, checks on indiscriminate developmental projects, better EIAs so that such consequences can be avoided in future.

6. Summary

India is an extremely influential actor in global environmental parleys because of her various identities like as an emerging economy, as a part of G-77 and China collective and as a member of BASIC group. Along with other emerging countries, she is vehemently protesting mandatory emissions cuts at various global fora. India's contribution to the present climate change is miniscule. During the period 1870–2019, it adds only 4 per cent to the global total emissions. In 2019, it was castigated as the world's third highest polluter, but its scale of emissions is 2.88 CO_2 gigatonnes (Gt) which is in no way comparable to that of China's (10.6 Gt) and the USA's (at 5 Gt).[131] However, its per capita emissions are also very low in comparison to the United States, the EU, the United Kingdom and China. In order to achieve the net-zero target by 2050, India has to adopt ambitious steps. In a projection made by the CSE, India would generate 4.48 Gt in 2030 under business-as-usual scenario. On per capita basis, India will be 2.31 tonnes per capita by that period. However, if this amount is compared with that of other big emitters, it would be found that the United States will be 9.42 tonnes, EU at 4.12 tonnes, the United Kingdom at 2.7 and China will be 8.88 CO_2 tonnes per capita in 2030.[132]

Question therefore remains that whether India needs to do more. The answer is India needs to curb emission for its own interest and it has already taken various steps in this direction. In recent years, India has become more proactive in climate negotiations as is evident since the Copenhagen. In order to maintain the position of a responsible actor in the global climate regime as well as an independent sovereign entity, she has made various strategic alliances which are guided by diplomatic equations. Since 2015, the Modi Government initiated strategic alliance with the US Government at the environmental front in a more rigorous way. Both countries concluded negotiations on a 5-year MoU on 'energy security, clean energy and climate change'.[133] In 2021, both the countries have again joined hands to cooperation at the climate front which was evident at the Leaders' Summit on Climate on 22 April 2021. A high-level India–US bilateral Partnership had been agreed to be launched in order to meet the goals of the Paris Agreement. The 'India-US Climate and Clean Energy Agenda 2030 Partnership' might herald a new era in the energy sector as well whose spillover effect will generate positive move to counter climate change. Both the countries have set ambitious targets as part of the new nationally determined contribution also. All of these would contribute towards strengthening bilateral collaboration over climate and clean energy.[134]

The post Kyoto world has also witnessed the strategically significant solidarity between India and China despite their traditional animosity and territorial disputes. They have been cooperating with each other in the climate negotiation field in order to strengthen their bargaining position vis-a-vis the developed world. Since the inception of Kyoto Protocol such alliances are evident between India and China. They together even strongly opposed the Kyoto provision of 'allowing a non-participating nation to choose, at any time and on a voluntary basis, a level of emissions control it felt was appropriate to its circumstances'.[135] It was only because of their resistance that the provision was struck out from the text. Not only that they have signed several MOUs on climate change cooperation, on green technology in order to explore low-carbon technology options fulfilling their rising energy demands. In October 2009, both the countries signed a 5-year 'Agreement on Cooperation on Addressing Climate Change' and promised to continue their coordination in the international climate negotiations. The agreement by providing an alternative framework helped the Indo-Chinese nexus to withstand pressure from the northern countries especially by the United States and Europe to take up mandatory emission cuts targets.[136] They rejected squarely any outside mandates that would hamper economic growth.

Along with this bilateral move, before Copenhagen both had announced unilateral voluntary commitment to reducing the carbon intensity of their development by 40–45 per cent from 2005 levels by 2020 for China and 20–25 per cent over the same time period for India.[137] However, India should not be treated at par with China in the climate negotiations as her emissions are well below China's per capita emissions. China is even far ahead of India in terms of total GDP and with reference to her carbon intensity of economy. Therefore, the hyphenated relation between India and China on the climate front may be counterproductive for India. Actually, in various climate summits especially at Copenhagen, China used India and other developing countries as the shield. Indian position is also not beyond criticism. Since the inception of climate talks, although the equity and poverty eradication have phrased her negotiating position, she is criticized for 'hiding behind the poor' while putting minimum efforts to establish equity and justice domestically. The CoP 26 that took place in Glasgow in 2021 once again witnessed Indo-China cooperation on climate front. Together they went against the pledge to phase out coal and petitioned to change the language to 'phase down' the use of coal. Ultimately, the term phase down had been omitted in the final agreement.[138] However, the picture is different in the developing block where India is losing faith among her natural allies as it was voicing its concern for CBDR and endorsing the per capita line of thinking in various CoPs. The recent India's effort to change the language of Glasgow agreement has fueled this mistrust again.

The natural resources of India are also affected by the market forces for which environmental harms have become more acute and have given birth to instabilities affecting social cohesion. It is a fact that agricultural sector of India is exposed to trade liberalization induced structural change for which some regions and farmers have received new opportunities of producing crops for export market and are benefiting from inflows of technology and investments, while others are facing undue competition from agricultural imports. Moreover, the multinational corporations have hijacked and monopolized the supply of seeds, thereby attacking the supply of indigenous seeds and community-based systems of food safety. The traditional wisdom and customs are threatened in the face of rapid use of genetic engineering, trawler fishing and commoditization. The environmental activist Vandana Shiva in her various writing has criticized the industrial model of agriculture which is a byproduct of the Green Revolution and is used to get more food. She has analysed in 'Stolen Harvest: The Hijacking of Global Food Supply' that diverse sources of food have been destroyed by industrial agriculture. Large quantity of food had been stolen from other species in order to bring market-oriented specific commodities. Such commodities use more fossil fuels and toxic chemical that act at the detriments of land and its produce.[139] We know that environmental change has adversely affected the natural capital of the country. But human actions and crave for more food, more growth actually exacerbated the resource scarcity. In India, many forest lands are destroyed and converted for monoculture of pine and eucalyptus that are used as industrial raw material to generate revenues and growth. However, it leads to biodiversity loss, and the most craved growth results in the destruction of the sources of food, fodder, fuel and medicine of the forest community and also has made them more vulnerable to floods and droughts. The unsustainable shrimp farming along the long coastlines of India has also results in devastation of both the coastal ecology and the livelihoods of coastal communities. Such commercial shrimp farming for the Western consumers has greatly affected the mangroves and gradually the coastal environment as the large shrimp trawlers kill species living in the sea bed. Thus, natural resources are diminishing causing detriments mostly to the poorer sections. Climate change affects these resources, but the corporate control and intervention mostly in the agricultural sector has worsened the situation and has created a green imperialism by which the farmers are turned into consumers of 'corporate patented products' and local and national markets are replaced by global ones.[140] As a result of this apartheid against nature, the poorer sections of the country and the farmers in particular have started lacking their right to food and life. Water resources and forests are also not free from the ill-effects of market forces. But Indian response to environmental change and its impact on these resources does not include any measure that prevents the commercialization of water or industrial interference in the forestlands. In both cases, the poorer sections are affected as though commercialization of these resources is aimed to augment the supply of them, but the benefits are reaped by the rich only.

Energy and climate change agenda has also taken centre stage in India's policy arena. The mammoth task of reducing global emissions needs re-orientation of our energy strategy that is based on renewable energy sources. Though India is following a massive drive in the production of renewables, it should be more aware of the fact that coal use and fossil fuel subsidies must be phased out. Fossil fuel based energy production may affect her other natural resources too. Like the thermal power plants require huge land and water, thereby affecting the availability of these resources for agriculture. The import of fossil fuels especially oil not only has made the country vulnerable to oil price spikes but hampers the energy security as well. Additionally, carbon-intensive goods may face international trade sanctions in future or tariffs may be levied on these goods. This may have significant impact on India's total export. Thus, as economic interest is closely related with managing the effects of environmental change, trade–energy efficiency–climate change nexus is very significant. The Indian policymakers should concentrate on this aspect while making energy policy as well.

From the previous analysis, it follows that there exist many issues that are related with environmental degradation and climate change which determines individual country's response to the threat. India is no exception. Her economic interests, human security issues and diplomatic interests are intertwined with her battle against environmental change. At the international level, her climate policy has largely been shaped by her diplomatic strategies while the domestic environmental policies have been guided by the need to maintain development and economic growth. As a result, our national responses to the problem and their representation at the global level are often guided by these priorities. But this process to remain oblivious to environmental concerns will stretch ourselves towards self-destruction. Managing natural resources in a sustainable way helps in achieving not only long-term economic interests but also conducive to mitigation and adaptation to climate change. Against such a backdrop, concerns for changing climate are inextricably linked with our national interest as India is largely vulnerable to climate disasters, and its economy is greatly dependent on climate-sensitive sectors. However, the ill-effects of globalization and the robbing of the natural capitals of the country in different ways may hinder her response to the issue of environmental change. Therefore, until and unless these issues are not properly addressed at both the national and global levels, no responsible environmental action, be it in the form of accepting mandatory emissions cuts or using sustainable methods for producing food or managing water and energy resources, will become a reality.

Notes

1 Indira Gandhi, *Of Man and His Environment*, Abhinav Publication: New Delhi, 1992, p.19.
2 *The Indian Wildlife Protection Act*, 1972. Available at envfor.nic.in/legis/wildlife/wildlife1.htm Accessed on December 27, 2013.
3 Somnath Roy, "Constitution of India and Environment Protection", in Sanjay Kumar Singh (ed.), *Environment Law and Climate Change*, SBS Publishers: New Delhi, 2010, pp.237–239.
4 Caesar Roy, "Emerging Trend in Environmental Law and the Contribution of Judiciary in India", in Sanjay Kumar Singh (ed.), *Environment Law and Climate Change*, SBS Publishers: New Delhi, 2010, pp.254–259.
5 The name of the Ministry of Environment and Forests has been changed to Ministry of Environment, Forests and Climate Change by the Narendra Modi government of India which aims to project India a responsible actor in international forums on climate change very forcefully.
6 Caesar Roy, No.4, pp.254–259.
7 Gautam Gupta, "Towards Responsible Environmentalism: The Global Order and the Case of India", in *Benefits of Environmental Policy: Conference Volume of the 6 Chemnitz Symposium: Europe and Environment*, Routledge: London, 2009, pp.89–91.
8 *The Environment (Protection) Act 1986*. Available at http://www.moef.nic.in/sites/default/files/eprotect_act_1986.pdf Accessed on December 27, 2013.
9 Hayley Stevenson, "India and International Norms of Climate Governance: A Constructivist Analysis of Normative Congruence", *Review of International Studies*, Vol.37, 2011, p.1014.
10 Sebastian Oberthur and Hermann E. Ott, *The Kyoto Protocol: International Climate Policy for the 21st Century*, Springer: Berlin, 1999, p.47.
11 *Energy Data Directory Yearbook*, TERI: New Delhi, 2009, p.532.
12 Sanjiv Kumar Sinha, "Climate Change Regime and Indian Approach", in Sanjay Kumar Singh (ed.), *Environment Law and Climate Change*, SBS Publishers: New Delhi, 2010, pp.102–103.
13 Government of India, "*India's First National Communication to the UNFCCC*", 2004, p.32. Available at http://unfccc.int/resource/docs/natc/indnc1.pdf Accessed on January 8, 2014.
14 Sandeep Sengupta, "International Climate Negotiations and India's Role", in Navroz K. Dubash (ed.), *Handbook of Climate Change and India: Development, Politics and Governance*, Oxford University Press: New Delhi, pp.108–109.
15 *Climate Change-India's Perspective*, Reference Note. No.25/RN/Ref./August/2013, p.9, Lok Sabha Secretariat, Parliament Library and Reference, Research, Documentation and Information Service: New Delhi, 2013.
16 C.P. Geevan, "National Environment Policy: Ascendance of Economic Factors", *Economic and Political Weekly*, October 23, 2004, pp.4686–4687.
17 Praful Bidwai, *The Politics of Climate Change: Mortgaging Our Future*, Orient Blackswan: New Delhi, 2012, p.85.
18 Eight National Missions are National Solar Mission, National Mission on Enhanced Energy Efficiency, National Water Mission, National Mission on Sustainable Habitat, National Mission for Sustaining the Himalayan Ecosystem, National Mission for Sustainable Agriculture and, National Mission on Strategic Knowledge for Climate Change.
19 Government of India, Prime Minister's Council on Climate Change, *National Action Plan on Climate Change*. Available at http://pmindia.nic.in/climate_change_english.pdf Accessed on January 16, 2014.

20 N.R. Krishnan, "The Climate Turned against India at Durban", *The Hindu*, December 13, 2011. Available at http://www.thehindubusinessline.com/opinion/the-climate-turned-against-india-at-durban/article2709519.ece Accessed on January 13, 2013.

21 Padmaparna Ghosh, "I Want to Position India as a Proactive Player: Jairam Ramesh", *Mint*, September 29, 2009. Available at http://www.livemint.com/Politics/h97ogi3qEaToXuP9T8YTuI/I-want-to-position-India-as-a-proactive-playerJairamRames.h Accessed on January 17, 2014.

22 Lok Sabha, *Transcript of Minister's Response in the Lok Sabha*, Parliament of India: New Delhi, December 3, 2009, p.246. Available at http://164.100.47.132/debatestext/15/III/0312.pdf Accessed on January 18, 2014.

23 Ibid., pp.238–239.

24 Ibid., p.246.

25 Sandeep Sengupta, No.14. pp.106–107.

26 N.H. Rabindranath, "The Copenhagen Accord", *Current Science*, Vol.98, No.6, March 25, 2010, p.752.

27 R.V. Anuradha, "Trade Law: The Road from Copenhagen", *The Times of India*, December 17, 2009.

28 Ibid.

29 Shikha Bhasin, Tobias Engelmeier and Felix Schimdt, "After Cancún India's New Role as an International Deal Maker", *Friedrich Ebert Stifung, Perspective*, April 2011, pp.2–6. Available at http://library.fes.de/pdf-files/iez/08145.pdf Accessed on March 21, 2015.

30 Lavanya Rajamani, "The Can-Can't at Cancun", December 14, 2010, *Indian Express*. Available at http://www.indian express.com/story-print/724374/ Accessed on January 14, 2014.

31 D. Raghunandan, "Durban Platform: Kyoto Negotiations Redux", *Economic and Political Weekly*, Vol.XLVI, No.53, December 31, 2011, p.15.

32 Lavanya Rajamani, "Deconstructing Durban", *The Indian Express*, December 15, 2011. Available at archive.indianexpress.com/news/deconstructing-durban/887892/ Accessed on January 28, 2014.

33 T. Jayaraman, "Post-Durban, India Has Its Task Cut Out", *The Hindu*, December 22, 2011.

34 PTI, "Include More Adaptation Efforts in Paris Climate Deal: India", *The Hindu*, December 3, 2014.

35 Meena Menon, "India to Press for Green Fund Operationalisation at Lima Climate Meet", *The Hindu*, November 30, 2014.

36 NRDC, *The Road from Paris-India's Progress Towards Climate Pledge*, Issue Brief. Available at https://www.nrdc.org/resources/road-paris-indias-progress-towards-its-climate-pledge Accessed on January 24, 2022.

37 Neil Macfarquhar, "U.N. Deadlock on Addressing Climate Shift", *New York Times*, July 20, 2011. Available at http://www.nytimes.com/2011/07/21/world/21nations.html?_r=0 Accessed on July 8, 2014.

38 A part of Section 2 previously published in Das, Satabdi, "Negotiating an Intractable Climate Deal: The Kyoto Process and Beyond", *Jadavpur Journal of International Relation*, December 1, 2013. doi.org/10.1177/0973598414535061

39 The Energy and Resources Institute, *Coping with Global Change: Vulnerability and Adaptation in Indian Agriculture*, 2003, p.23. Available at http://www.iisd.org/pdf/2004/climate_coping_global_change.pdf Accessed on July 11, 2014.

40 Department of Agriculture and Cooperation, Ministry of Agriculture, Government of India, *National Food Security Mission*, August 2007. Available at http://agricoop.nic.in/nfsm/nfsm.pdf Accessed on July 10, 2014.

41 Rajeswari Raina, "Agriculture in the Environment: Are Sustainable Climate Friendly Systems Possible in India?", in Navroz K. Dubash (ed.), *Handbook of Climate Change and India: Development, Politics and Governance*, Oxford University Press: New Delhi, 2012, pp.317–319.

42 Ministry of Agriculture, Department of Agriculture and Cooperation, Government of India, *National Mission for Sustainable Agriculture: Strategies for Meeting the Challenges of Climate Change*, August 2010. Available at http://www.indiaenvironmentportal.org.in/files/file/National%20Mission%20For%20 Sustainable%20Agriculture.pdf Accessed on July 10, 2014.

43 Rajeshwari Raina, No.40, pp.317–321.

44 Vandana Shiva, "Monsanto and the Mustard Seeds", *Earth Island Journal*. Available at http://www.earthisland.org/journal/index.php/eij/article/monsanto_and_the_mustard_seed/ Accessed on February 14, 2015.

45 Vandana Shiva, *People Should See That Corporations Have Abandoned Them Long Ago*. Available at https://www. youtube.com/watch?v=6GE6o9Z4sQ8 Accessed on July 13, 2014.

46 Praful Bidwai, No.17, p.150.

47 Mahim Pratap Singh, "Fertilizer Subsidy Hurts Everyone", *The Hindu*, July 10, 2014.

48 Rajeswari Raina, No.40, pp.320–321.

49 K.S. Kavi Kumar, Priya Shyamsundar and A. Arivudai Nambi, "Economics of Climate Change Adaptation in India", *Economic and Political Weekly*, Vol.XLV, No.18, May 1, 2010, p.27.

50 Greenpeace, *Ensuring Our Food Security*. Available at http://www.greenpeace. org/india/en/What-We-Do/Sustainable-Agriculture/ Accessed on July 13, 2014.

51 *In Farmers' Protests, a Climate Change Connection Lurks*, December 10, 2020. Available at In Farmers' Protests, a Climate Change Connection Lurks – The Wire Science, Accessed on January 23, 2022.

52 *National Action Programme to Combat Desertification*. Available at http://envfor.nic.in/public-information/national-action-programme-combat-desertification-0 Accessed on November 19, 2014.

53 Dennis Taenzler, Lukas Ruettinger, Katherina Ziegenhagen (adelphi) and Gopalakrishna Murthy, *Water, Crisis and Climate Change in India: A policy Brief, The Initiative for Peacebuilding – Early Warning Analysis to Action* (IfP-EW), October 2011, p.10. Available at www.ifp-ew.eu/pdf/201110IfPEWW aterCrisisIndiaPolBrf.pdf Accessed on July 18, 2014.

54 Philippe Cullet, *Water Law in India: Overview of Existing Framework and Proposed Reforms*, International Environmental Law Research Centre, 2007, p.5. Available at http://www.ielrc.org/content/w0701.pdf Accessed on July 17, 2014.

55 Food and Agricultural Organization, *India Aquastat*, 2010, p.15. Available at http://www.fao.org/nr/water/aquastat/countries_regions/IND/IND-CP_eng.pdf Accessed on July 18, 2014.

56 Asian Development Bank, *Water Resources Development in India: Critical Issues and Strategic Options*, March 1, 2009. Available at http://www.indiaenvironmentportal.org.in/content/269510/water-resources-development-in-india-critical-issues-and-strategic-options/ Accessed on July 17, 2014.

57 Himanshu Kulkarni and Himanshu Thakkar, "Framework for India's Strategic Water Resource Management Under a Changing Climate", in Navroz K. Dubash (ed.), *Handbook of Climate Change and India: Development, Politics and Governance*, Oxford University Press: New Delhi, 2012, p.337.

58 Planning Commission, Government of India, Report of the Expert Group, *Groundwater Management and Ownership*, September 2007, p.25. Available at planningcommission.gov.in/reports/genrep/rep_grndwat.pdf Accessed on July 18, 2014.

59 Ministry of Water Resources, Government of India, National Water Mission to the National Action Plan on Climate Change, Vol.1, April 2009. Available at www.moef.nic.in/downloads/others/Mission-SAPCC-NWM.pdf Accessed on July 18, 2014.
60 Praful Bidwai, No.17, pp.147–148.
61 "Water Sector Needs Purposeful Action to Match India's Long Term Needs", *The Economic Times*, March 29, 2014.
62 Ministry of Water Resources, Government of India, *Draft National Water Policy*, 2012, pp.9–10. Available at www.downtoearth.org.in/dte/userfiles/.../ DraftNWP2012.p... Accessed on July 19, 2014.
63 Prashant Prabhakar Deshpande, "National Water Policy and Action Plan for India 2020 – Part 2", *The India Times*, November 8, 2021. Available at india-times.com Accessed on January 24, 2022.
64 Gautam Gupta, "Towards Responsible Environmentalism: The Global Order and the Case of India", in *Benefits of Environmental Policy: Conference Volume of the 6 Chemnitz Symposium: EU and Environment*, Routledge: London, 2009, p.90.
65 Ministry of Environment and Forests, "Phase 2 of Ganga Action Plan", Ministry of Environment and Forests, August 9, 2011. Available at http://pib.nic.in/ newsite/erelease.aspx?relid=74173 Accessed on July 19, 2014.
66 Ramaswamy R. Iyer, "For Rejuveneting, Not Reengineering, the Ganga", *The Hindu*, July 16, 2014.
67 V.S. Vyas and V. Ratna Reddy, "Assessment of Environmental Policies and Policy Implementation in India", *Economic and Political Weekly*, January 10, 1998, p.51.
68 Ridhima Sud, Jitendra Vir Sharma, Arun Kumar Bansal and Subhash Chandra, *Institutional Framework for Implementing REDD+ in India*, Ministry of Environment and Forest, Government of India, The Energy and Resources Institute, Norwegian Embassy, 2012. Available at www.indiaenvironmentportal.org.in/.../ institutional-framework-for-imple... Accessed on July 20, 2014.
69 The Ministry of Environment and Forests, Government of India and Food and Agricultural Organizations to the United Nations Regional Office for Asia and the Pacific, Working Paper No. APFSOS II/WP/2009/06, India Forestry Outlook Study, 2009, pp.16–18. Available at www.fao.org/docrep/014/am251e/ am251e00.pdf Accessed on July 12, 2013.
70 N.H. Ravindranath, Rajiv Kumar Chaturvedi and Indu K. Murthy, "Forest Conservation, Afforestation and Reforestation in India: Implications for Forest Carbon Stocks", *Current Science*, Vol.95, No.2, July 25, 2008, pp.217–219.
71 Ministry of Environment and Forests, Government of India, "Sustainable Development in India: Stocktaking in the Run Up to Rio+20", 2011, pp.17–18. Available at envfor.nic.in/sites/default/files/Sust_Dev_Stocktaking_0.pdf Accessed on July 23, 2014.
72 P.R. Shukla, Subodh Kumar Sharma, Amit Garg, Sumana Bhattacharya and N.H. Rabindranath, "Vulnerability and Adaptation Challenge Ahead", in R. Shukla, Subodh Kumar Sharma, Amit Garg, Sumana Bhattacharya and N.H. Rabindranath (eds.), *Climate Change and India: Vulnerability Assessment and Adaptation*, University Press: Hyderabad, 2003, p.412.
73 The Ministry of Environment and Forest, Government of India, and Food and Agricultural Organizations to the United Nations Regional Office For Asia and the Pacific, Working Paper No. APFSOS II/WP/2009/06, India Forestry Outlook Study, 2009, pp.16–18. Available at www.fao.org/docrep/014/am251e/ am251e00.pdf Accessed on July 12, 2013.
74 Shankar Gopalkrishnan, "'Mitigation or Exploitation: The Climate Talks, REDD and Forest Areas", in Navroz K. Dubash (ed.), *Handbook of Climate*

Change and India: Development, Politics, and Governance, Oxford University Press: New Delhi, 2012, pp.346–347.

75 The Ministry of Environment and Forests, Government of India and Food and Agricultural Organizations to the United Nations Regional Office for Asia and the Pacific, Working Paper No. APFSOS II/WP/2009/06, India Forestry Outlook Study, 2009, p.16.

76 Ridhima Sud, Jitendra Vir Sharma, Arun Kumar Bansal and Subhash Chandra, No.67.

77 Dr J.V. Sharma and Priyanka Kohli, *Forest governance and implementation of REDD+ in India*, Ministry of Environment and Forest, Government of India and TERI. Available at indiaenvironmentportal.org.in/.../forest-governance-and-implementation... Accessed on July 20, 2014.

78 Arnab Kumar Hazra, *History of Conflict over Forests in India: A Market Based Resolution*, Working Paper Series, Julian L. Simon Centre for Policy Research, April 2002, pp.7–9. Available at www.libertyindia.org/policy_reports/forest_conflict_2002.pdf Accessed on July 16, 2014.

79 "Is JFM Relevant?", *Down to Earth*, September 15, 2011. Available at http://www.downtoearth.org.in/content/jfm-relevant Accessed on July 20, 2014.

80 Duncan Brack and Katharina Umpfenbach, "Deforestation and Climate Change", *The World Today*, October 2009, p.9.

81 Ministry of Environment and Forest, Government of India, India's Forests and REDD+". Available at www.moef.nic.in/downloads/public-information/REDD-report.pdf Accessed on July 23, 2014.

82 Ridhima Sud, Jitendra Vir Sharma, Arun Kumar Bansal and Subhash Chandra, No.67.

83 Shankar Gopalkrishnan, No.73. pp.348–350.

84 Vandana Shiva, *Staying Alive: Women, Ecology and Survival in India*, Zed Books: London, 1988, pp.81–82.

85 Vandana Shiva, *Water Wars: Privatization, Pollution and Profit*, South End Press: Cambridge, 2002, pp.4–5.

86 Shankar Gopalkrishnan, No.73, pp.348–350.

87 Ministry of Environment and Forests, Government of India National Mission for a Green India (Under The National Action Plan on Climate Change) Draft submitted to Prime Minister's Council on Climate Change, p.4. Available at www.moef.nic.in/downloads/public.../GIM-Report-PMCCC.pdf Accessed on June 6, 2013.

88 Praful Bidwai, *The Politics of Climate Change and the Global Crisis: Mortgaging Our Future*, Orient Blackswan: Hyderabad, p.156.

89 Ridhima Sud, Jitendra Vir Sharma, Arun Kumar Bansal and Subhash Chandra, No.67.

90 Himanshu Thakkar, *Water Sector Options for India in a Changing Climate*, South Asia Network on Dams, Rivers & People: New Delhi, March 2012, p.9. Available at sandrp.in/wtrsect/Ex_Summary_WATER_SECTOR_OPTIONS_FOR_IN... Accessed on July 23, 2014.

91 Praful Bidwai, No.17, p.157.

92 The Ministry of Environment and Forests Government of India and Food and Agricultural Organizations to the United Nations Regional Office for Asia and the Pacific, Working Paper No. APFSOS II/WP/2009/06, India Forestry Outlook Study, 2009, pp.18–20.

93 A part of Section 3.3 previously published in an edited book, Das, Satabdi, "Participatory Sustainable Forest Resource Management: The role of Joint Forest Management in India", in K.C. Lalmalsawmzauva et al. (eds.), *Natural Resource Management for Sustainable Development*, 2019. Today & Tomorrow's Printers and Publishers: New Delhi.

94 "UN Meet on Biodiversity", *The Statesman*, October 9, 2012.

95 "End Biopiracy", *Deccan Herald*, October 10, 2012.

96 Jyoti Parikh, "India's Efforts to Minimize Greenhouse Gas Emission: Policies, Measures and Institution", in Velma I. Grover (ed.), *Global Warming and Climate Change: Ten Years after Kyoto and Still Counting*, Vol.1, Science Publishers: Enfield, NH, 2008, pp.333–339.

97 Ministry of Power, Bureau of Energy Efficiency, Government of India, *National Mission for Enhanced Energy Efficiency, Draft Mission Document: Implementation Framework*, 2008. Available at www.indiaenvironmentportal. org.in/.../ national-mission-for-enhanced-en... Accessed on July 27, 2014.

98 Kirit S. Parikh, "India's Options to Meet Energy Requirement", in Jyotirmay Mathur, H.J. Wagner and N.K. Bansal (eds.), *Energy Security, Climate Change and Sustainable Development*, Anamaya Publisher: New Delhi, 2007, pp.12–13.

99 Planning Commission, Government of India, New Delhi, *Integrated Energy Policy: Report of the Expert Committee*, August 2006. Available at http:// planningcommission.nic.in/reports/genrep/rep_intengy.pdf Accessed on July 29, 2014.

100 P. Balachandra, Darshini Ravindranath and N.H. Ravindranath, "Energy Efficiency in India: Assessing the Policy Regimes and Their Impacts", *Energy Policy*, Vol.38, No.11, November 2010, p.6433.

101 Saran Shyam, *India's Climate Change Policy: Towards a Better Future*, November 8, 2019. Available at https://www.mea.gov.in/articles-in-indian-media.htm? dtl/32018/Indias_Climate_Change_Policy_Towards_a_Better_Future Accessed on January 21, 2022.

102 Prateek Bumb and Rituraj, *Carbon Dioxide Capture and Storage (CCS) in Geological Formations as Clean Development Mechanism (CDM) Projects Activities (SBSTA)*. Available at cdm.unfccc.int/about/ccs/docs/CCS_geo.pdf Accessed on July 29, 2010.

103 Vijay Joshi and Urjit R. Patel, "India and a Carbon Deal", *Economic and Political Weekly*, Vol.XLIV, No.31, August 1, 2009, pp.74–76.

104 Dharmmala Nikhilesh, "The Role of Carbon Capture and Storage in India's 'Hard to Abate' Industries", *The Wire*, February 19, 2021. Available at https://science.thewire.in/economy/energy/what-role-does-carbon-capture-and-storage-play-in-indias-climate-puzzle/ Accessed on January 24, 2022.

105 Roger Harrabin, "Indian Farm Makes Carbon Capture Breakthrough", *The Guardian*, January 4, 2017. Available at https://www.theguardian. com/environment/2017/jan/03/indian-firm-carbon-capture-breakthrough carbonclean#:~:text=A%20breakthrough%20in%20the%20race,it%20 to%20make%20baking%20soda Accessed on January 12, 2022.

106 Global Energy Network Institute, *Overview of Sustainable Renewable Energy Potential of India*, January 2010. Available at http://www.geni.org/ globalenergy/research/renewable-energy-potential-of-india/Renewab%20 Energy%20Potential%20for%20India.pdf Accessed on September 8, 2013.

107 Shebonti Ray Dadwal, "India's Renewable Energy Challenge", *Strategic Analysis*, Vol.34, No.1, January 2010, p.2.

108 Mahim Pratap Singh, "Rajasthan Tops in Solar Energy", *The Hindu*, May 10, 2013.

109 Ministry of New and Renewable Energy, Government of India, *Jawaharlal Nehru National Solar Mission*. Available at http://www.mnre.gov.in/solar-mission/jnnsm/introduction-2/ Accessed on July 29, 2014.

110 Narendra Kumar Tripathi, "Hydropower in Asia: Spinning a Dependence and Interdependence Binary", *South Asian Survey*, Vol.17, No.2, September 2010, pp.222–223.

111 ICLEI South Asia, *Renewable Energy and Energy Efficiency Status in India*, May 2007, p.15. Available at http://localrenewables.iclei.org/fileadmin/template/projects/localrenewables/files/Local_Renewables/Publications/RE_EE_report_India_final_sm.pdf Accessed on September 8, 2013, p.68.

112 Girjesh Pant, "The Future of Energy Security", *South Asian Survey*, Vol.17, No.1, 2010, p.40.

113 J. Venkatesan, "Fear of Nuclear Disaster Has No Basis", *The Hindu*, May 7, 2013.

114 Sunita Narain, "India's New Climate Targets: Bold, Ambitious and a Challenge for the World", *Down to Earth*, November 2021. Available at https://www.downtoearth.org.in/blog/climate-change/india-s-new-climate-targets-bold-ambitious-and-a-challenge-for-the-world-80022 Accessed on January 24, 2022.

115 Robert J. Nicholl, "Planning for the Impacts of Sea Level Rise", *Oceanography*, Vol.24, No.2, 2011, p.144.

116 Ministry of Environment and Forest, *Notification Under Section 3(1) and Section 3(2)(v) of the Environment (Protection) Act and Rule 5(3)(d) of the Environment (Protection) Rules, 1986 Declaring Coastal Stretches as Coastal Regulation Zone (CRZ) and Regulating Activities in the CRZ*, 1991. Available at http://envfor.nic.in/legis/crz/crznew.html Accessed on February 17, 2015.

117 Ministry of Environment and Forest, Annexure-1, *Coastal Area Classification and Development Regulations*, 1991. Available at http://envfor.nic.in/legis/crz/crznew.html Accessed on February 17, 2015.

118 Sumana Narayanan, "New Rules for Coasts", *Down to Earth*, February 15, 2011. Available at http://www.downtoearth.org.in/content/new-rules-coasts Accessed on February 17, 2015.

119 "New Coastal Management Rules", *The Hindu*, June 2, 2008.

120 World Environment Day, Press Release, 2014. Available at www.moef.nic.in/sites/default/files/Presss-release-WED-2014_0.pdf Accessed on July 31, 2014.

121 Ibid.

122 Suman Chakraborti and Rohit Khanna, "World Bank May Take Back Coastal Funds", *The Times of India*, June 11, 2013.

123 Srestha Banerjee, "Green Tribunal Halts Construction on Puri Coast", *Down to Earth*, December 7, 2013. Available at http://www.downtoearth.org.in/content/green-tribunal-halts-construction-puri-coast Accessed on July 31, 2014.

124 The Government of India, Ministry of Home Affairs, Disaster Management in India, 2005. Available at www.unisdr.org/2005/mdgs-drr/national-reports/India-report.pdf Accessed on August 1, 2014.

125 Press Information Bureau, Government of India National Policy on Disaster Management (NPDM). Available at http://pib.nic.in/newsite/erelease.aspx?relid=53359 Accessed on August 1, 2014.

126 Sambhav S. Kumar, "How Effective Is India's Disaster Management Authority?", *Down to Earth*, June 20, 2013. Available at http://www.downtoearth.org.in/content/how-effective-indias-disaster-management-authority Accessed on August 1, 2014.

127 National Disaster Management Authority, *Snapshot on Efforts on Community Based Disaster Management in India*. Available at http://www.ndma.gov.in/en/get-involved/community-based-disaster-management/introduction-citi.html Accessed on August 1, 2014.

128 National Policy on Disaster Management (NPDM), *Press Information Bureau*, Government of India. Available at http://pib.nic.in/newsite/erelease.aspx?relid=53359 Accessed on August 1, 2014.

129 Ministry of Science and Technology, Press Information Bureau, Government of India, *Approval for the National Mission for Sustaining Himalayan*

Ecosystem Launched under the National Action Plan on Climate Change. Available at http://pib.nic.in/newsite/PrintRelease.aspx?relid=104353 Accessed on August 1, 2014.

130 Ministry of Science and Technology, Government of India, *Mission Document of National Mission for Sustaining the Himalayan Eco System,* June 2010, pp.1–4. Available at http://dst.gov.in/scientific-programme/NMSHE_June_2010.pdf Accessed on August 1, 2010.

131 Sunita Narain, "India's New Climate Targets: Bold, Ambitious and a Challenge for the World", *Down to Earth*, November 2, 2021. Available at downtoearth. org.in Accessed on January 12, 2022.

132 Ibid.

133 Vishwa Mohon, "Obama-Modi Climate Deal: Unlike China, No Emission Target for India", *Times News Network*, January 26, 2015. Available at http://economictimes.indiatimes.com/news/environment/global-warming/ obama-modi-climate-deal-unlike-china-no-emission-target-for-india/article-show/46016298.cms Accessed on February 13, 2015.

134 MEA, Government of India, India-US Joint Statement on Launching the "India-US Climate and Clean Energy Agenda 2030 Partnership", April 22, 2021. Available at mea.gov.in Accessed on January 24, 2022.

135 Henry D. Jacoby, Prinn G. Ronald and Richard Schmalensee, "Kyoto's Unfinished Business", *Foreign Affairs*, Vol.77, No.4, July–August 1998, p.63.

136 William R. Hawkins, "China – India Accord to Scuttle UN Climate Treaty", *American Thinker*, October 23, 2009. Available at www.americanthinker. com/2009/10/chinaindia_accord_to_scuttle_u.html Accessed on March 19, 2014.

137 Toufiq Siddiqi, "China and India: More Cooperation Than Competition in Energy and Climate Change", *Journal of International Affairs*, Fall/Winter, Vol.64, No.1, 2010, p.78.

138 William Gittins, *COP26: What Was India and China's Last Minute Change to the Climate Crisis Agreement?* November 13, 2021. Available at AS.com Accessed on January 24, 2022.

139 Vandana Shiva, *Stolen Harvest: The Hijacking of Global Food Supply*, South End Press: Cambridge, MA, 2000, p.12.

140 Ibid., pp.1–8.

Concluding Remarks

Today, it has become more compelling with scientific evidence that environment is changing in response to global warming. The impact of these changes in environment includes the physical effects like droughts, floods and rising severity of natural disasters and their resultant fluctuations in the supply of key resources such as disruption in water availability, poor harvest, declining food stocks, loss of biodiversity and so on. All of these have amplified the fragility of people and communities leading to rise in numbers of migrants and internally displaced persons. Though these impacts of environmental change can be felt over longer time and there exists uncertainty regarding the scale and geographical spread of the impacts, the global community have realized that the cost of inaction is larger than actions.

Inevitable Consequences of Environmental Change Pose Security Risks to Countries

Environmental change is an immediate risk to a nation's security. It affects the natural capital of a country, destroys lives and property, thereby affecting the economy, growth and development. It leaves countries living on the edge more vulnerable to climate catastrophes and resultant socio- economic disruptions while intensifying competition among nations vying for limited resources. Restrictions in the availability of life-sustaining resources are inherent in all societies but marginalized ones are the worst sufferer. Actually, environmental stress as a threat amplifier interacts with their politico-economic and social landscapes where the existence of weak state institutions, unstable political arrangements and consequent armed and violent conflicts are very common. The end result is a metamorphosis in the status of environmental threat from a mere scientific issue to a major component of a country's domestic policy formulations and foreign policy designs. Environmental security has thus entered the realm of the discourse of present security studies. Given the all-pervasive and intergenerational impacts of environmental problems, it has been expected that countries will act in unison to combat the threat. But as vulnerability, coping capacity and resource distribution differ across regions, countries' responses also

DOI: 10.4324/9781003271192-6

differ. The way in which the global community has negotiated its response to the threat is highly politicized resulting in clashes of interests between the actors. Though the style of living of the affluent North and their earlier depredation of nature is responsible for the danger of environmental change mostly, the rising economies are equally responsible for the catastrophe of their present status of unsustainable consumption.

The commoditization of the atmosphere is also critical. The idea of extending property rights to the environment along with trading of these rights through market like international trading of carbon emissions in the form of payments for ecosystem services (e.g. REDD) or pricing and trading carbon emissions reduction (through CDM) gives birth to 'carbon colonialism'[1] where the terms of trade are determined by the global North. In this process of market intervention in the arena of public goods and services and the retreat of government from this sphere, the goal of investors, companies and traders to make profit out of environmental goods precedes the urge to halt climate change. The developing countries are suffering the most as their emissions are rising for generating carbon credits and the developed world has also exported polluting industries and their byproducts to them. Another point of serious discontent includes financial support and technology transfer that the developing countries need to curb emissions. While the developed countries are in favour of market-driven financial contribution, the developing countries vehemently protest against that. The promise of financial assistance also remains largely inoperative. Moreover, because of profit generation, most of the climate finances are used for emission-cutting projects in middle-income countries. However, the amount needed for adaptation in poor countries is minuscule. Even funding for loss and damage are also neglected, while the ravages of climate crisis prevent countries' adaptation process. In case of technology transfer, the exploitation of the Southern countries prevail as developed countries are keener on making profit out of this technology transfer and in maximizing royalty earnings.

The impacts of environmental change and their associated consequences are thus threats to individual, national and global security. They have affected the economic growth, the socio-political stability of a country on the one hand, while at the global level, contributed to political and diplomatic tussles based on divisive national interests of countries, thereby minimizing the scope of cooperation. All of these have created a new agenda in security studies that include different actors and stakeholders other than the government of a state. Governments are not the sole drivers and promoters of this securitization process where the referent object is the environment. There are other non-state actors like supranational bodies, civil society groups and NGOs who are continuously securitizing the environment through 'speech acts' or through the repeated discussion of environmental threats in the public domain, thereby transforming the perceived environmental problems into matters of national and international security concerns. Such environmental problems have become more acute as social and economic inequalities are inherent to them. Additionally, responses to the environmental risks have also revealed the hierarchical and institutional structure of decision-making both within a given society and at the global environmental parleys, which may generate further insecurities.

Environmental Insecurity Has Socio-Political and Economic Origins in India

Environmental issues are related with economy, social and political setup of India. Here, the scarcities of natural resources are sometimes human constructs and augmented by lack of will and inefficiency of governments to distribute or mobilize resources in a just way among communities and due to lack of decentralization in the decision-making process regarding the protection of such resources. The poor sections are mostly affected by them and the latter have no other options at their disposal but to use the available natural capital in an unsustainable way. The problem has further intensified due to the privatization and commercialization of these natural resources. The corporate control and intervention mostly in the agricultural sector has contributed to the green imperialism by which the real harvesters have become consumers of 'corporate patented products' and local and national markets are replaced by global ones. A kind of fabricated scarcity results from this. Farmers have no other options but to buy seeds at high prices and are caught in the debt trap. Sometimes privatization has been used to maintain the supply of natural goods like water where also the rural poor are the greatest sufferer. In view of such energy poverty across India, the Modi government stressed on deploying more eco-friendly energy measures and investment in the renewables mainly in the solar and wind power sectors. The aim is to make a 'paradigm shift' from the 'carbon credit' based on emissions cut to 'green credit'. Prime Minister Modi has made commitments several times to ensure universal energy access for India's poor as without that the fruits of development cannot be reaped by all equally. Though in the energy sector the government has made such significant breakthrough, it has been criticized by various quarters due to its actions taken time to time affecting the lives and livelihood securities of forest-dependent communities. One such instance was the central government notification underlining rules for exemption of projects in the forest lands having no tribal population according to the recent census. The cut-off date for considering the forest land that is subject to conversion for such project was 13 December 1930. Ironically considering this date most of the forest lands might be used for corporate purpose. Such move had been criticized from all quarters. The Ministry of Environment, Forest and Climate Change (MoEFCC) order dated 28 October 2014, also denied the previous practice of seeking permission from the village council before converting a forest land for industrial purposes.[2] All of these are instanced as the government's attempt to undermine the rights of the forest-based rural communities and to cause harm to the carbon sinks of India.

Urbanization and industrialization process of the government in the name of economic growth have thus also furthered environmental decay. Unscrupulous developmental projects, mining in geo-ecologically fragile zones with manipulated environmental impact assessment for fulfilling narrow economic gains of certain classes, have contributed to this. The people living at the edge are paying the cost of these catastrophes which are the results of human greed. Here, the politico-social and economic marginalization in India is interwoven with the environmental miseries.

India's National Interest Clashes with Global Concern for Climate Change

Undoubtedly India is suffering from environmental insecurities. As an emerging economy, it is a crucial actor in the global environmental regime. She has underscored the notion of equity in the environmental change debate by proclaiming the principle of CBDR initially. In various multilateral policy fora, India on behalf of the developing countries is asserting socio-economic development and poverty eradication as her primary policy objectives. She has considered the climate mitigation option as an impediment in achieving these goals. It is therefore in her national interest to seek for greater carbon space that is required for fulfilling the country's developmental aspirations. This is also the reason behind the per capita emission line of thinking which is inherent to her climate negotiation position.

India has often been criticized for her 'feet dragging' climate posture by the developed world, and it is argued that she is hiding behind the poor while showing little eagerness to follow a clean growth and climate resilient trajectory that require relatively little ecological space. But, starting from Paris climate summit of 2015 to Glasgow summit of 2021, there are significant changes in India's climate policy. In the post-2020 international climate regime, all countries are asked to significantly increase efforts for emissions reduction while simultaneously asked to increase resilience to climate change impacts. Against such a backdrop, in the Glasgow Summit, India had made several ambitious targets to reach the net-zero emission which in a way is a shift in India's hard-line environmental position that any agreement should apply only to the developed world. However, Indo-China opposition to the commitment to phasing out coal while negotiating the final agreement again created disarray within India's natural constituency-G-77 and Indian posture had been criticized by some developed countries as well. India's national interest here comes in conflict with the global concerns in two ways. First, question remains regarding the different capabilities of the countries given their different stages of development. India's efforts to meet the 1.5°C target must be viewed through the lens of its efforts to eradicate poverty. Removing coal and subsidy in fossil fuel entirely might come in conflict with the country's sustainable development goal. Second, while India is expected to switch over from fossil fuels to renewable, did developed countries fulfil the assured financial help by 2020 and technology transfer. Despite the fact India's track record with renewables in recent times is quite laudable, according to the CEA, in 2019, renewables contribute more than 9 per cent of electricity generation in India. In 2021, the percentage had been increased to 12 per cent as there is a huge increase in renewables (102 GW).[3] Even the energy plan of India as stated reveals her positive move to restrict its coal-based energy.

Each and every stakeholder therefore has its own priorities while devising climate strategies at the global level. India is not an exception. She has significant stakes in the global environmental parleys. But her goal is to have an equitable share of the available carbon space. This space is required for fulfilling her developmental aspirations. It is also true at the same time that embarking on cleaner developmental path is advantageous for India as rising emissions and resultant environmental change due to 'dirty' developmental policies are harming her population too. But it is not altogether possible for India to immediately reduce atmospheric release of GHG emissions drastically by a sudden transition to less carbon-intensive economy. So, though global warming and environmental degradation are attracting global concerns to a large extent, on the global platform individual country's interest clashes with it on various grounds. In the same manner, India's economic interest and developmental aspirations have a clash with the process and outcomes of global environmental negotiations. But the country has made repeated attempts to take significant steps for combating environmental threats without jeopardizing her national interest.

Sustainable Development Is a Possible Means to Bridge the Gulf Between Development and Environmental Protection in India

India and other developing countries have prioritized development over environmental protection in various environmental summits. But there are gaps in integrating the goals of economic and social development and environmental sustainability. Poverty itself is a constraint in achieving adaptive capacity to cope with changing climate. So the aim for environmental protection must be clubbed with poverty reduction and sustainable development efforts. For this, 'equitable access to sustainable development' is required that depends on equitable share of the available carbon space. Therefore, equity in burden sharing and equitable apportion of global 'carbon budget' are cardinal issues in India's climate posture. Its demands include to bridge the gulf between entitlement-wise share in the global carbon space and the actual physical availability of the same that can be accessed by the country. She is well aware of the fact that due to the over-occupation of global carbon space by the developed North, it is hard to get the fair share of global carbon budget. However, it is also true that being an emerging economy she also needs to reduce her emissions subsequently. So, equity in accessibility to available carbon space must be circumscribed by the sustainability criterion which requires appropriate financial and technology transfer from the developed countries. Such transfer may help India and other developing countries to address their requirement for environmentally sound technologies. India has therefore mooted the idea of sustainable development fund and has projected on behalf of developing block the need to make technologies both accessible and affordable at various summits.

However, in reality, India and other developing countries have to buy these technologies at exorbitant rate and these technologies are often incompatible with the local systems. The traditional knowledge base that is potential enough to foster sustainability is often ignored by the governments of developing countries. The Indigenous communities who live mostly at the 'ecological margins' of human settlement are the finest observers of changes in environment. Their traditional ways of living, dependence on biodiversity and ecological services, can provide rich insights for sustainable living and well-being. But these communities and their knowledge and expertise regarding sustainable resource management and development are not considered during environmental policymaking at all levels. The community-based mitigation and adaptation actions are not taken into considerations in global environmental deals. Some measures like JFM, REDD+ though have endeavoured to involve Indigenous people in managing resources like forestlands, they have often undermined the customary rights of Indigenous people upon natural resources. Often the frameworks of these resource management policies are exteriorly designed with minimum transparency and Indigenous people usually find no representation in decision-making, thereby leaving the process open to manipulations.

Another mechanism facilitating sustainable developments in developing countries is the CDM. Though it is essential for promoting sustainable development in countries that host CDM projects, India's approach to this is not satisfactory. Unlike China she does not steer CDM investment in tandem with her policy priorities. Mostly CDM projects are concentrated in large industrialized states like Gujarat and Maharashtra, whereas less-developed states have few such projects in comparison to their populations like in Bihar and Uttar Pradesh. So in case of sustainable development, India faces dual problems – first, Indigenous knowledge is not used properly and she has to depend on western technologies and funding that do not encourage innovation of low-cost options for bringing sustainability and accommodate local mitigation and adaptation approaches that can generate employment as well. Second, she neither promotes nor discourages the implementation of CDM projects in various states and does not invest in capacity building for implementing these projects in small, less-developed states. As a result they lack CDM created investment opportunities. Such differences in allotting projects have escalated the rich poor dichotomy again which is inherent in Indian society. The end result is balanced sustainable development which is necessary for combating environmental threats fails to become a reality.

The global environmental politics has involved various stakeholders who may differ in their opinion about the intensity of the problem, determining the degree of vulnerability, devising the ways to combat the threat as well as regarding the matter of securitizing the environmental issue. India is not an exception. Her response to environmental degradation and climate change is inextricably linked with her national interest which may not always converge with the global concerns for the threat. She is responding to this crisis in her own way. Her national interest is therefore not to lose the developmental opportunities in various multilateral negotiations and regimes while responding to global environmental issues.

At the global level, India is voicing her concern for equity in terms of per capita emissions in the hope that it would provide her with more carbon space to fulfil the developmental needs. In the post-2020 climate regime, the concept of equity has been discarded as it might create obstacles in raising and reaching higher ambitions for mitigation. Respective capabilities are underscored dynamically with reference to CBDR and equity principles now. However, this is a contradiction in spirit as countries' stages of development are not the same, and capability wise they also differ. The right to development of each country must not be neglected as it ensures sustainable development and the livelihood security for the poor multitudes. The bottom line of equity principle articulates this. So the other associated issues like finance and technology transfer all must be seen through the prism of equity. Despite promising several ambitious mitigation targets, India has demanded that in the post-2020 climate regime, equity must be an integral part of all arrangements.

Equity must be established at all levels. Within India there exists inequity in resource distribution, in the availability of socio-political and economic safety nets at the time of environmental disasters and people of all strata do not find representations in the policymaking. The development process is also not all-encompassing engaging everybody's participation. The livelihood security of the impoverished is rarely maintained in a hierarchical Indian society. Here, the level of overall emissions is rising due to the luxurious emissions of the affluent which is further escalated by the survival emissions of the poor multitude. The emancipation of India's poor masses needs the integration of environmental concerns with the socio-economic development process, but partial policy formulations failed to achieve this goal. Environmental security should be established at the national level first. Until those most vulnerable are capable of achieving a healthy and productive life which is environmentally sustainable too, any effort to raise voice for securing India's national interest, that is, balancing the goal of environmental protection with her developmental imperatives, will be unsuccessful at the global platform.

Notes

1 The term coined by Larry Lohman, founder of the Durban Group for Climate Justice.
2 Basudev Mahapatra, *The Modi Government's War on Environment*, November 12, 2014. Available at https://www.opendemocracy.net/openindia/basudev-maha patra/modi-government%E2%80%99s-war-on-environment Accessed on March 13, 2015.
3 Sunita Narain, "India's New Climate Targets: Bold, Ambitious and a Challenge for the World", *Down to Earth*, November 2, 2021. Available at https://www. downtoearth.org.in/blog/climate-change/india-s-new-climate-targets-bold-ambitious-and-a-challenge-for-the-world-80022 Accessed on January 24, 2022.

Bibliography

Reports and Documents

Asian Development Bank, *Asian Development Outlook 2019: Strengthening Disaster Resilience*. Available at https://www.adb.org/sites/default/files/publication/492711/ado2019.pdf Accessed on November 13, 2019.

Asian Development Bank, *Water Resources Development in India: Critical Issues and Strategic Options*, March 1, 2009. Available at http://www.indiaenvironmentportal.org.in/content/269510/water-resources-development-in-india-critical-issues-and-strategic-options/ Accessed on July 17, 2014.

Australian Government, Department of Environment, *Cartagena Dialogue for Progressive Action, 2010–2011*. Available at http://www.climatechange.gov.au/about-us/annual-reports/annual-report-2010-11/feature-cartagena-dialogue-progressive-action Accessed on March 19, 2014.

Center for Climate and Energy Solution, *CoP 12 Report Twelfth Session of the Conference of the Parties to the UN Framework Convention on Climate Change and Second Meeting of the Parties to the Kyoto Protocol"*, 2006. Available at www.c2es.org/international/negotiations/cop-12/summary Accessed on June 6, 2014.

Centre for Education and Documentation, *Forests and Climate Change in India*. Available at http://base.d-p-h.info/en/fiches/dph/fiche-dph-8613.html Accessed on July 15, 2013.

Climate and Development Knowledge Network, *Agriculture and Climate Change Policy Brief: Main Issues for UNFCCC and Beyond*. Meridian Institute, 2011. Available at www.climate-agricul ture.org/~/media/Files/.../ACCPolicy_web.pdf Accessed on May 16, 2014.

Climate Change-India's Perspective (Reference Note. No. 25/RN/Ref.), August 2013, Lok Sabha Secretariat, Parliament Library and Reference, Research, Documentation and Information Service: New Delhi.

Climate Change, Migration and Displacement, *Who Will Be Affected? Working Paper Submitted by the Informal Group on Migration/Displacement and Climate Change of the IASC*, October 31, 2008. Available at http://unfccc.int/resource/docs/2008/smsn/igo/022.pdf Accessed on 16 June, 2014.

Convention Relating to the Status of Refugees, Art. 1, July 28, 1951. Available at http://www.unhcr.org/3b66c2aa10.pdf Accessed on December, 25, 2011.

Decision of 31 January 1992 (3046th Security Council Meeting): Statement by the President, *Consideration of Questions Under the Responsibility of the Security Council in the Maintenance of International Peace and Security, Chapter VIII*,

Repertoire of the Practice of the Security Council, 1989–1992. Available at http://www.un.org/en/sc/repertoire/89-92/Chapter%208/GENERAL%20ISSUES/Item%2028_SC%20respons%20in%20maint%20IP S.pdf Accessed on June 28, 2014.

Declaration of Human Rights, 1948. Available at http://www.un.org/en/documents/udhr/ Accessed on June 17, 2014.

Definition of Food Security Given by FAO at World Food Summit 1996. Available at http://www.who.int/trade/glossary/story028/en/ Accessed on December 20, 2014.

Department of Agriculture and Cooperation, Ministry of Agriculture, Government of India, *National Food Security Mission*, August 2007. Available at http://agricoop.nic.in/nfsm/nfsm.pdf Accessed on July 10, 2014.

Desertification, Drought and Climate Change, Africa Report, 2008. Available at http://www.un.org/esa/sustdev/publications/trends_africa2008/desertification.pdf Accessed on October 20, 2014.

Desertification: Environmental Degradation. Available at http://www.un.org/en/events/desertificationday/background.shtml Accessed on November 15, 2014.

European Commission, *EU at COP 26 Climate Change Conference.* Available at https://ec.europa.eu/info/strategy/priorities-2019-2024/european-green-deal/climate-action-and-green-deal/eu-cop26-climate-change-conference_en Accessed on December 13, 2021.

FAO, *A Guide to Agriculture at UNFCCC COP 19*, November 11–22, 2013. Available at http://www.fao.org/docrep/019/ar716e/ar716e.pdf Accessed on January 22, 2015.

FAO, *Climate Change and Food Security: A Framework Document*, 2008. Available at www.fao.org/forestry/15538-079b31d45081fe9c3dbc6ff34de 4807e4.pdf Accessed on May 19, 2014.

FAO, *Climate Change, Water and Food Security, Technical Background Document From the Expert Consultation*, February 26–28, 2008. Available at http://www.fao.org/docrep/016/ap526e/ap526e.pdf Accessed on February 21, 2015.

FAO, *India Aquastat*, 2010. Available at http://www.fao.org/nr/water/aquastat/countriesregions/IND/IND-CP_eng.pdf Accessed on July 18, 2014.

Final Chairman's Summary: First Major Economies Meeting on Energy Security and Climate Change, White House Council on Environmental Quality, September 27–28, 2007. Available at http://2001-2009.state.gov/g/oes/climate/mem/93021.htm Accessed on June 24, 2014.

Forests: Climate Change, Biodiversity and Land Degradation, Brochure Published by the Joint Liaison Group of the Rio Convention, 2008. Available at http://www.cbd.int/doc/publications/for-cc-2008-en.pdf Accessed on December 21, 2014.

Gender CC-Women for Climate Justice, *Briefing Paper on UNFCCC and Agriculture*, 2012. Available at www.gendercc.net/uploads/media/BriefingPaper_Agriculture 2012.pdf Accessed on May 19, 2014.

German Advisory Council on Global Change, *Climate Change as a Security Risk*, Earthscan: London, 2009.

Gleneagles Plan of Action on Climate Change, Clean Energy and Sustainable Development, July 2005. Available at https://www.gov.uk/government/uploads/system/uploads/attachmentdata/file/48584/glenea gles-planofaction.pdf Accessed on June 24, 2014.

Government of Bihar, *World Bank Global Facility for Disaster Reduction & Recovery, Bihar Kosi Flood: Needs Assessment Report*, 2008. Available at http://www.

gfdrr.org/sites/gfdrr.org/files/publication/GFDRR_India_PDNA_2010_EN.pdf)
Accessed on August 28, 2013.

IDMC, *Global Report on Internal Displacement*, 2021. Available at https://www.
internal-displacement.org/global-report/grid2021/ Accessed on December 23,
2021.

IDSA Task Force Report, *Water Security for India: The External Dynamics*, Institute
for Defence Studies and Analyses: New Delhi, 2010.

Internal Displacement Monitoring Center and Norwegian Refugee Council, *Global
Estimates 2012: People Displaced by Disasters*, May 2013. Available at http://
www.downtoearth.org.in/themes/DTE/htm/global-estimates-2012-may2013.pdf,
Accessed on August 28, 2013.

International Covenant on Civil and Political Rights. Adopted by the General
Assembly of the United Nations on December 19, 1966. Available at http://www.
refworld.org/pdfid/3ae6b3aa0.pdf Accessed on June 17, 2014.

International Energy Agency, *Progress with Implementing Energy Efficiency Policies
in the G8*, 2009. Available at http://www.iea.org/publications/freepublications/
publication/G8Energy efficiencyprogressreport.pdf Accessed on June 24, 2014.

International Energy Agency, *World Energy Outlook*, 2006. Available at http://
www.worldenergyoutlook.org/media/weowebsite/2008-1994/WEO2006.pdf
Accessed on June 23, 2014.

International Energy Agency, *World Energy Outlook, 2007: China and India Insights*.
Available at http://www.worldenergyoutlook.org/media/weowebsite/2008-1994/
weo2007.pdf Accessed on December 30, 2014.

International Organisation for Migration, *Compendium of IOM's Activities, Migration, Climate Change and the Environment*, 2009. Available at http://www.iom.
int/jahia/webdav/shared/shared/mainsite/activities/envdegradation/compendium-
climatechange.pdf Accessed on December 21, 2014.

International Organisation for Migration, *International Dialogue on Migration,
No.18, Climate Change, Environmental Degradation and Migration*, 2012. Available at www.iom.int/.../workshops/clim Accessed on June 17, 2014.

International Organisation for Migration, *Migration, Climate Change and the Environment*, Policy Brief, May 2009. Available at http://www.iom.int Accessed on
December 21, 2014.

IPCC, *Analysing Regional Aspects of Climate Change and Water Resources*, Chapter 5. Available at www.ipcc.ch/pdf/technical-papers/ccw/chapter5.pdf Accessed
on June 12, 2013.

IPCC, *Introduction to Climate Change and Water*, June 2008. Available at http://
www.ipcc.ch/pdf/technical-papers/ccw/chapter1.pdf Accessed on June 2, 2014.

Kimberley Declaration, 2002. Available at www.tebtebba.org/.../17-rio-10-world-
summit-on-sustainable-d... Accessed on June 10, 2014.

*Letter Dated 16 April 2007 from the Permanent Representative of Pakistan to the
United Nations Addressed to the President of the Security Council*. Available at
http://www.securitycouncilreport.org/atf/cf/%7B65BFCF9B-6D27-4E9C-8CD3-
CF6E4FF96FF9%7D/CC%20S2007%20211.pdf Accessed on June 29, 2014.

MEA, Government of India, *India-US Joint Statement on Launching the "India-
US Climate and Clean Energy Agenda 2030 Partnership"*, April 22, 2021.
Available at India-US Joint Statement on Launching the "India-US Climate and
Clean Energy Agenda 2030 Partnership" (mea.gov.in) Accessed on January 24,
2022.

Ministry of Agriculture, Department of Agriculture and Cooperation, Government of India, *National Mission for Sustainable Agriculture: Strategies for Meeting the Challenges of Climate Change*, August 2010. Available at http://www.indiaenvironmentportal.org.in/files/file/National%20Mission%20For%20Sustaina ble%20Agriculture.pdf Accessed on July 10, 2014.

Ministry of Environment and Forest, Government of India, *Executive Summary of India's Third National Report to Convention on Biological Diversity*, 2006. Available at https://www.cbd.int/doc/world/in/in-nr-03-p1-en.pdf Accessed on July 26, 2014.

Ministry of Environment and Forest, Government of India, *India's Fifth National Report to the Convention on Biological Diversity*, 2014. Available at https://www.cbd.int/doc/world/in/in-nr-05-en.pdf Accessed on November 14, 2014.

Ministry of Environment and Forests, *Government of India and Food and Agricultural Organizations to the United Nations Regional Office for Asia and the Pacific, Working Paper No. APFSOS II/WP/2009/06, India Forestry Outlook Study*, 2009. Available at www.fao.org/docrep/014/am251e/am251e00.pdf Accessed on July 12, 2013.

Ministry of Environment and Forests, Government of India, *India's First National Communication to the UNFCCC*, 2004. Available at http://unfccc.int/resource/docs/natc/indnc1.pdf Accessed on January 8, 2014.

Ministry of Environment and Forests, Government of India, *India's Initial National Communication to the UNFCCC, National Communication (NATCOM), Vulnerability Assessment and Adaptation*, 2004. Available at unfccc.int/resource/docs/natc/indnc1.pdf Accessed on February 21, 2015.

Ministry of Environment and Forests, Government of India, *India's Forests and REDD+*, November 24, 2010. Available at www.moef.nic.in/downloads/public-information/REDD-report.pdf Accessed on July 23, 2014.

Ministry of Environment and Forests, Government of India, *Indian Network for Climate Change Assessment, India: Greenhouse Gas Emissions 2007*, May 2010. Available at http://moef.nic.in/sites/default/files/EXECUTIVE%20SUMMARY-PS+HRP.pdf Accessed on September 5, 2013.

Ministry of Environment and Forests, Government of India, *India's Second National Communication to the United Nations Framework Convention on Climate Change*, 2012. Available at http://envfor.nic.in/downloads/public-information/India%20Second%20National%20Communication%20to%20UNFCCC.pdf Accessed on July 26, 2013.

Ministry of Environment and Forests, Government of India, *National Mission for a Green India (Under the National Action Plan on Climate Change)*, Draft submitted to Prime Minister's Council on Climate Change, 2008. Available at www.moef.nic.in/downloads/public.../GIM-Report-PMCCC.pdf Accessed on June 6, 2013.

Ministry of Environment and Forests, Government of India, *Phase 2 of Ganga Action Plan*, August 9, 2011. Available at http://pib.nic.in/newsite/erelease.aspx?relid=74173 Accessed on July 19, 2014.

Ministry of Environment and Forests, Government of India, *Sustainable Development in India: Stocktaking in the Run Up to Rio+20*, 2011. Available at envfor.nic.in/sites/default/files/Sust_Dev_Stocktaking_0.pdf Accessed on July 23, 2014.

Ministry of Environment and Forests, *Indian Network for Climate Change Assessment, Climate Change and India: A 4X4 Assessment, Sectoral and Regional*

Analysis for 2030, November 2010. Available at http://www.moef.nic.in/down loads/public-information/fin-rpt-incca.pdf Accessed on August 1, 2013.

Ministry of External Affairs, *Documents, PM's Intervention on Climate Change at the Heiligendamm Meeting*, June 8, 2007. Available at http://www.mea.gov.in/ in-focus-article.htm?18822/PMs+intervention+on+Climate+Change+at+the+Heil igendamm+meeting Accessed on January 13, 2013.

Ministry of Home Affairs, *The Government of India, Disaster Management in India*, 2005. Available at www.unisdr.org/2005/mdgs-drr/national-reports/India-report.pdf Accessed on August 1, 2014.

Ministry of Law and Justice, Government of India, *the Gazette of India, The Disaster Management Act*, 2005. Available at http://nafoindia.org/pdfs/the_disaster_ management_act_2005.pdf Accessed on August 1, 2014.

Ministry of New and Renewable Energy, Government of India, *Jawaharlal Nehru National Solar Mission*. Available at http://www.mnre.gov.in/solar-mission/jnnsm/ introduction-2/ Accessed on July 29, 2014.

Ministry of Power, Bureau of Energy Efficiency, Government of India, *National Mission for Enhanced Energy Efficiency, Draft Mission Document: Implementation Framework*, 2008. Available at www.indiaenvironmentportal.org.in/.../national-mission-for-enhanced-en... Accessed on July 27, 2014.

Ministry of Power, Government of India, *National Mission on Enhanced Energy Efficiency: For Public Comments*, January 1, 2009. Available at http://www. indiaenvironmentportal.org.in/content/283989/national-mission-on-enhanced-energy-efficiency-for-public-comments Accessed on July 28, 2014.

Ministry of Power, Government of India, *National Mission on Enhanced Energy Efficiency NMEEE: Note*, August 1, 2009. Available at http://www. indiaenvironmentportal.org.in/content/284120/national-mission-on-enhanced-energy-efficiency-nmeee-note/ Accessed on July 29, 2014.

Ministry of Science and Technology, Government of India, *Mission Document of National Mission for Sustaining the Himalayan Eco System*, June 2010. Available at http://dst.gov.in/scientific-programme/NMSHE_June_2010.pdf Accessed on August 1, 2010.

Ministry of Water Resources, Government of India, *Draft National Water Policy*, 2012. Available at www.downtoearth.org.in/dte/userfiles/.../DraftNWP2012.p... Accessed on July 19, 2014.

Ministry of Water Resources, Government of India, *National Water Mission under National Action Plan on Climate Change, Comprehensive Mission Document, Volume – I*, April 2009. Available at www.nicra-icar.in/.../Mission%20Documents/ WATER%20MISSION.pdf Accessed on June 12, 2013.

Minority Rights Group International, *State of World's Minorities and Indigenous People*, 2012. Available at http://www.minorityrights.org/download.php@id= 1112 Accessed on February 15, 2015.

Nassau Forum Adopts Declaration, Strategy Paper in Preparation for Mauritius Meeting on Small Island States, DEV/2456, Press Release, January 30, 2004. Available at www.un.org/News/Press/docs/2004/dev2456.doc.htm Accessed on June 13, 2014.

National Disaster Management Authority, *Snapshot on Efforts on Community Based Disaster Management in India*. Available at http://www.ndma.gov.in/ en/get-involved/community-based-disaster-management/introduction-citi.html Accessed on August 1, 2014.

National Inputs of India for the Rio+20, 2012. Available at http://www.uncsd2012. org/content/documents/49NationalInputs_ofIndia_forRio20.pdf Accessed on April 24, 2014.

NRDC, *The Road from Paris-India's Progress Towards Climate Pledge*, Issue Brief. Available at NRDC: The Road from Paris – India's Progress Toward Its Climate Pledge (PDF) Accessed on January 24, 2022.

Outcome Document of Rio+20 Subregional Preparatory Meeting of SIDS of the Atlantic, Indian Ocean, Mediterranean and South China Sea (AIMS) Subregions, July 2011. www.uncsd2012.org/.../documents/AIMS%20Rio+20%20 Outcome%20d Accessed on June 14, 2014.

PEW Center on Global Climate Change, *Climate Change 101: International Action*, 2011. Available at http://www.c2es.org/docUploads/climate101-fullbook.pdf Accessed on January 8, 2014.

PEW Center on Global Climate Change, *Summary: Copenhagen Climate Summit*, 2009. Available at http://www.c2es.org/international/negotiations/cop-15/summary Accessed on June 6, 2014.

Planning Commission, Government of India, *Integrated Energy Policy, Report of the Expert Committee*, August 2006. Available at http://planningcommission.nic.in/ reports/genrep/intengpol.pdf Accessed on September 5, 2013.

Planning Commission, Government of India, *Report of the Expert Group, Groundwater Management and Ownership*, September 2007. Available at planningcommission.gov.in/reports/genrep/rep_grndwat.pdf Accessed on July 18, 2014.

Planning Commission, Government of India, *Water Resources*. Available at planning commission.nic.in/plans/.../chap21_water.... Accessed on July 19, 2014.

Potsdam Institute for Climate Impact Research and Climate Analytics, *Turn Down the Heat: Climate Extremes, Regional Impacts and the Case for Resilience, A Report for the World Bank*, June 2013. Available at http://www.worldbank. org/content/dam/Worldbank/document/Full_Report_Vol_2_Turn_Down_The-Heat_%20Climate_Extremes_Regional_Impacts_Case_for_Resilience_Print%20 version_FINAL.pdf Accessed on July 24, 2013.

Press Information Bureau, Government of India, *National Policy on Disaster Management (NPDM)*. Available at http://pib.nic.in/newsite/erelease.aspx?relid=53359 Accessed on August 1, 2014.

Prime Ministers' Council on Climate Change. Available at http://pmindia.nic.in/ committeeescouncils details.php?nodeid=7 Accessed on January 11, 2014.

Prime Minister's Statement at the Commonwealth Summit, November 2009. Available at http://meaindia.nic.in/speech/2009//11/27/ss01.htm Accessed on January 14, 2014.

Promoting Sustainable Agriculture and Rural Development, *Chapter 14, Agenda 21*, 1992. Available at http://www.fao.org/sd/erp/toolkit/Books/SARD LEARNING/CD-SL/SourcesAgenda%2021-chapter2014.htm Accessed on June 21, 2014.

Proposal by the Alliance of Small Island States (AOSIS) for the Survival of the Kyoto Protocol and a Copenhagen Protocol, December 11, 2009. Available at http://www.indiaenvironmentportal.org.in/content/293517/proposal-by-the-alliance-of-small-island-states-aosis-for-the-survival-of-the-kyoto-protocol-and-a-copenhagen-protocol/ Accessed on June 14, 2014.

Protection of the Quality and Supply of Freshwater Resources: Application of Integrated Approaches to the Development, Management and Use of Water Resources,

Chapter 18, Agenda 21, 1992. Available at http://www.earthsummit2002.org/ic/freshwater/reschapt18b.html Accessed on May 27, 2014.

Quito Declaration: Recommendations of Indigenous Peoples and Organizations Regarding the Process of the Framework Convention on Climate Change, 2000. Available at http://www.wrm.org.uy/oldsite/actors/CCC/Quito.html Accessed on June 9, 2014.

Security Council Open Debate: UK Concept Paper, Energy, Security and Climate, 2007. Available at https://unfccc.int/files/application/pdf/ukpaper_securitycouncil.pdf Accessed on June 28, 2014.

Statement by Mr. Shyam Saran, Special Envoy of PM on Climate Change, at the Closing Plenary Session of AWG-LCA of UNFCCC Climate Change Meeting, November 6, 2009. Available at http://meaindia. nic.in/speech/2009//11/06/ss03.htm Accessed on January 14, 2014.

Statistical Review of World Energy, 2021. Available at Full report – Statistical Review of World Energy 2021 (bp.com) Accessed on January 22, 2022.

Submission by the Governments of Papua New Guinea and Costa Rica to the Eleventh Conference of the Parties to the UNFCCC: Agenda Item 6, Reducing Emissions from Deforestation in Developing Countries: Approaches to Stimulate Action, December 2005. Available at rainforestcoalition.org/.../COP-11AgendaItem6-Misc.Doc.FINAL.pdf Accessed on June 10, 2014.

TERI, *India: Climate-friendly Development, Published for Ministry of Environment and Forest, Government of India,* IG Printers: New Delhi, 2002. Available at http://moef.nic.in/sites/default/files/cc/cop8/moefbk/moefbk.pdf Accessed on January 7, 2014.

Text of the UNFCCC. Available at http://unfccc.int/essential_background/convention/background/items/1353.php Accessed on May 16, 2014.

The Dublin Statement on Water and Sustainable Development, 1992. Available at www.un-documents.net/h2o-dub.htm Accessed on May 27, 2014.

The Energy and Resources Institute, *Coping with Global Change: Vulnerability and Adaptation in Indian Agriculture,* 2003. Available at http://www.iisd.org/pdf/2004/climate_coping_global_change.pdf Accessed on July 11, 2014.

The Energy and Resources Institute, *Looking Back to Change Track: Green India 2047.* Available at www. Icsudev.org/summary_greenindia.pdf. Accessed on January 5, 2013.

The Environment (Protection) Act, 1986. Available at http://www.moef.nic.in/sites/default/files/eprot ectact_1986.pdf Accessed on December 27, 2013.

The General Assembly Resolution, 63/281, *Climate Change and Its Security Implications,* 2009. Available at http://www.un.org/esa/dsd/resources/res_pdfs/ga-64/cc-inputs/IcelandCCIS.pdf Accessed on June 28, 2014.

The Indian Wildlife Protection Act, 1972. Available at envfor.nic.in/legis/wildlife/wildlife1.htm Accessed on December 27, 2013.

The UN General Assembly, *2005 World Summit Outcome, 16th Session,* September, 2005. Available at www.un.org/womenwatch/ods/A-RES-60-1-E.pdf Accessed on June 23, 2014.

The UN General Assembly, *Mauritius Strategy for the Further Implementation of the Programme of Action for the Sustainable Development of Small Island Developing States, Report of the Secretary General,* October 3, 2005. Available at http://unctad.org/en/Docs/a60d401en.pdf Accessed on June 13, 2014.

The United Nations Convention to Combat Desertification, Part-1, Use of Terms, 1994. Available at http://www.unccd.int/en/about-the-convention/Pages/Text-Part-I.aspx Accessed on November 17, 2014.

The White House, Office of the Press Secretary, *Declaration of the Leaders of the Major Economic Forum on Energy and Climate*, July 9, 2009. Available at http://www.whitehouse.gov/the_press_office/Declaration-of-the-Leaders-the-Major-Economies-Forum-on-Energy-and-Climate Accessed on 14 January, 2014.

The World Bank, Agriculture and Rural Development Unit, South Asian Region, *Report No. 34750-IN, India's Water Economy: Bracing for a Turbulent Future*, December 22, 2005. Available at www-wds.worldbank.org/external/default/.../WDSP/IB/.../34750.pdf Accessed on May 29, 2013.

The World Bank, *Fuel for Thought: An Environmental Strategy for the Energy Sector*, June 2000. Available at http://documents.worldbank.org/curated/en/2000/06/443544/fuel-thought-environmental-strategy-energy-sector Accessed on June 21, 2014.

Transcript of Minister's Response in the Lok Sabha, Parliament of India: New Delhi, December 3, 2009. Available at http://164.100.47.132/debatestext/15/III/0312.pdf Accessed on January 18, 2014.

UN REDD Programme, *The UN REDD Programme and REDD+*, November 2010. Available at http://www.unep.org/forests/Portals/142/docs/UN-REDD%20FAQs%20%5B11.10%5D.pdf Accessed on August 13, 2014.

UNDP, UNEP, UNESCAP, UNFCCC, UNISDR and WMO, *TST Issue Brief: Climate Change and Disaster Risk Reduction*, 2013. Available at http://sustainabledevelopment.un.org/content/documents/2301TST%20Issue%20Brief_CC&DRR_Final_4_Nov_final%20final.pdf Accessed on August 25, 2014.

UNEP, *Barbados Programme of Action for the Sustainable Development of Small Island Developing States*, 1994. Available at http://www.unep.ch/regionalseas/partners/sids.htm Accessed on June 12, 2014.

UNEP, *Emerging Issues for Small Island Developing States: Results of the UNEP Foresight Process*, June 2014. Available at http://www.indiaenvironmentportal.org.in/files/file/Emerging%20issues%20for%20small%20island%20developing%20states.pdf Accessed on June 12, 2014.

UNEP, *Integrating Environment and Development: 1972–2002*. Available at www.unep.org/GEO/geo3/pdfs/Chapter1.pdf Accessed on March 14, 2013.

UNEP, *One Planet and Many People: Atlas of Our Changing Environment*, 2006. Available at https://na.unep.net/atlas/onePlanetManyPeople/images/chapters/Atlas_Introduction_Screen.pdf Accessed on February 9, 2015.

UNEP, *The Importance of Mangroves to People: A Time to Action*, 2014. Available at http://www.indiaenvironmentportal.org.in/files/file/The%20importance%20of%20mangroves%20to%20people%20a%20call%20to%20action-2014.pdf Accessed on December 16, 2014.

UNFCCC, *Article 1.9*. Available at http://unfccc.int/essential_background/convention/background/items/1362txt.php Accessed on June 5, 2014.

UNFCCC, *Article 3.3*. Available at http://unfccc.int/essential_background/convention/background/items/1362txt.php Accessed on June 5, 2014.

UNFCCC, *Article 4.1(c) and 4.1(d)*. Available at http://unfccc.int/essential_background/convention/background/items/1362txt.php Accessed on June 5, 2014.

UNFCCC, *Fact Sheet: Reducing Emissions from Deforestation in Developing Countries: Approaches to Stimulate Action*, June 2009. Available at unfccc.int/.../backgrounders.../fact_ sheet_reducing_ emissions_from_def... Accessed on June 6, 2014.

UNFCCC, *Global Energy Interconnection Is Crucial for Paris Goals*, November 28, 2021. Available at https://unfccc.int/news/global-energy-interconnection-is-crucial-for-paris-goals Accessed on January 20, 2022.

UNFCCC, *Land Use, Land-Use Change and Forestry (LULUCF)*. Available at https://unfccc.int/methods/lulucf/items/3060.php Accessed on May 16, 2014.

UNFCCC, *National Adaptation Programmes of Action (NAPAs)*. Available at http://unfccc.int/national_reports/napa/items/2719.php Accessed on June 17, 2014.

UNFCCC, *Report of the Conference of Parties on Its Third Session held at Kyoto from 1–11 December*, 1998. Available at https://cdm.unfccc.int/Reference/COPMOP/08a01.pdf Accessed on March 14, 2013.

UNFCCC, *Report of the Conference of the Parties on its thirteenth session, held in Bali from 3–15 December 2007*, March 14, 2008. Available at unfccc.int/resource/docs/2007/cop13/eng/06a01.pdf Accessed on April 14, 2013.

UNFCCC, *Technical Paper, Challenges and Opportunities for Mitigation in the Agricultural Sector*, November 21, 2008. Available at http://unfccc.int/resource/docs/2008/tp/08.pdf Accessed on May 16, 2014.

United Nations Convention to Combat Desertification, 1994. Available at http://www.unccd.int/Lists/Site DocumentLibrary/convention/leaflet_eng.pdf Accessed on November 15, 2014.

United Nations Forum for Forests, *IPF/Iff Process (1995–2000)*. Available at http://www.un.org/esa/forests/ipf_iff.html Accessed on June 5, 2014.

UN-Water Thematic Initiative, *Coping with Water Scarcity*, August 2006. Available at www.un.org/waterforlifedecade/.../2006_unwater_coping_with_water_sc Accessed on May 28, 2014.

Vital Signs: The Trends that are Shaping our Future, Report prepared by the Worldwatch Institute: New York, 2007–2008.

World Commission on Environment and Development, *Our Common Future, Brundtland Report*, Oxford University Press: London, New York, 1987.

World Food Program, *Climate Change and Hunger: Towards a WFP Policy on Climate Change*, March 15, 2011. Available at documents.wfp.org/stellent/groups/public/documents/.../wfp 232740.pdf Accessed on May 19, 2014.

Yokohoma Strategy and Plan of Action, 1994. Available at http://www.unisdr.org/files/8241_doc6841contenido1.pdf Accessed on August 25, 2014.

Books

Achanta, Amrita N. (ed.), *The Climate Change Agenda: An Indian Perspective*, Tata Energy Research Institute: New Delhi, 1993.

Bakker, Karen, *Privatizing Water: Governance Failure and the World's Urban Water Crisis*, Orient Blackswan: Hyderabad, 2010.

Bandopadhyay, Jayanta, Kanchan Chopra and Nilanjan Ghosh (eds.), *Environmental Governance: Approaches, Imperatives and Methods*, Bloomsbury: London, 2012.

Bast, Joseph L. and Diane Carol Bast (eds.), *Climate Change Reconsidered: 2009 Report of the Nongovernmental International Panel on Climate Change*, The Heartland Institute: Arlington Heights, IL, 2009.

Benefits of Environmental Policy: Conference Volume of the 6 Chemnitz Symposium: Europe and Environment, Routledge: London, 2009.

Bhattacharya, Purusottam and Sugata Hazra, *Environment and Human Security*, Lancer's Book: New Delhi, 2003.

Bidwai, Praful, *An India That Can Say Yes: A Climate Responsible Development Agenda for Copenhagen and Beyond*, Heinrich Boll Stiftung: New Delhi, 2009.

Bidwai, Praful, *The Politics of Climate Change: Mortgaging Our Future*, Orient Blackswan: Hyderabad, 2012.

Bryant, Raymond L. and Sinead Bailey, *Third World Political Ecology*, Routledge: London, 1996.

Buzan, Barry, Ole Waever and Jaap de Wilde, *Security: A New Framework for Analysis*, Lynne Rienner: Boulder, CO, 1998.

Buzan, Barry, *People, State and Fear: The National Security Problem in International Relations*, Harvester Press: Great Britain, 1987.

Calvert, Peter and Susan Calvert, *The South, the North and the Environment*, Pinter: London and New York, 1999.

Chandran, D. Subha and J. Jeganaathan, *Energy and Environmental Security: A Cooperative Approach in South Asia*, Institute of Peace and Conflict Studies: New Delhi, 2011.

Chellaney, Brahma, *On the Frontline of Climate Change*, Konrad-Adenauer-Stiftung: New Delhi, 2007.

Chellaney, Brahma, *Water: Asia's New Battleground*, Georgetown University Press: Washington, DC, 2011.

Conca, Ken, *Governing Water: Contentious Transnational Politics and Global Institution Building*, The MIT Press: Cambridge, MA, 2006.

Dalby, Simon, *Environmental Security, Broadlines, Volume 20*, University of Minnesota Press: Minneapolis, MN and London, 2002.

Damodaran, A., *Encircling the Seamless: India, Climate Change and the Global Commons*, Oxford University Press: New Delhi, 2010.

Delmas, Magali A., and Oran R. Young, *Governance for Environment: New Perspectives*, Cambridge University Press: Cambridge, 2009.

Dobson, Andrew, *Justice and the Environment: Conceptions of Environmental Sustainability and Dimensions of Social Justice*, Oxford University Press: Oxford, 1998.

Dodds, Felix and Tim Pippard (eds.), *Human and Environmental Security: An Agenda for Change*, Earthscan: London, 2005.

Driessen, Paul, *Eco Imperialism: Green Power, Black Death*, Academic Foundation: New Delhi, 2005.

Dubash, Navroz K. (ed.), *Handbook of Climate Change and India: Development, Politics and Governance*, Oxford University Press: New Delhi, 2012.

Dunn, Myriam and Victor Mauer (eds.), *The Routledge Handbook of Security Studies*, Routledge: London, 2010.

Energy Data Directory Yearbook, TERI: New Delhi, 2009.

Gaan, Narottam, *Climate Change and International Politics*, Kalpaz Publications: New Delhi, 2008.

Gautam, Col. P.K., *Environmental Security: Internal and External Dimensions and Response*, A United Service Institution of India Project, DS Kothari Chair, Knowledge World: New Delhi, 2003.

Giddens, Anthony, *The Politics of Climate Change*, Polity Press: Cambridge, 2009.

Gradziuk, Artur and Ernest Wyciszkiewicz (eds.), *Energy Security and Climate Change: Double Challenge for Policy Makers*, The Polish Institute of International Affairs: Warsaws, 2009.

Grover, Velma I. (ed.), *Global Warming and Climate Change: Ten Years after Kyoto and Still Counting*, Volume 1, Science Publishers: Enfield, NH, 2008.

Gupta, Jayeeta and Nicolien Van Der Grijp, *Mainstreaming Climate Change in Development Cooperation: Theory, Practice and Implications for the Europe*, Cambridge University Press: Cambridge, 2010.

Gupta, Sujata and R.K. Pachauri (eds.), *Global Warming and Climate Change: Perspectives from Developing Countries*, The Energy and Resource Institute: New Delhi, 1989.

Harris, Paul G. (ed.), *The Politics of Climate Change: Environmental Dynamics in International Affairs*, Routledge: London, 2009.

Harrison, Kathryn and Lisa McIntosh Sundstorm, *Global Commons, Domestic Decisions: The Comparative Politics of Climate Change*, The MIT Press: Cambridge, MA, 2010.

Hecker, Jeanne Hyde, *Promoting Environmental Security and Poverty Alleviation in the Peat Swamps of Central Kalimantan, Indonesia*, Institute for Environmental Security: The Hague, 2005.

Houltart, Francois, *Agrofuels: Big Profits, Ruined Lives and Ecological Destruction*, Pluto Press: London, 2010.

Hulme, Mike, *Why We Disagree About Climate Change: Understanding Controversy, Inaction and Opportunity*, Cambridge University Press: Cambridge, 2009.

Hurrell, Andrew and Benedict Kingsburry, *The International Politics of the Environment: Actors, Interests and Institutions*, Clarendon Press: Oxford, 1992.

Iyer, Ramaswamy R., *Water: Perspectives, Issues, Concerns*, Sage Publications: New Delhi, 2003.

Jagmohon, Crisis of Environment and Climate Change, Allied Publishers: Mumbai, 2008.

Jayaram, Dhanasree, *Breaking Out of the Green House: Indian Leadership in Times of Environmental Change*, KW Publishing House: New Delhi, 2012.

Jones, Bruce, Carlos Pascual and Stephen John Stedman, *Power and Responsibility: Building International Order in an Era of Transnational Threats*, Brookings Institution Press: Washington, DC, 2009.

Kannan, A., *Global Environmental Governance and Desertification: A Study of Gulf Cooperation Council Countries*, Concept Publishing Company: New Delhi, 2012.

Karan, Pradyumna P., and Shanmugam P. Subbiah, *The Indian Ocean Tsunami: The Global Response to a Natural Disaster*, Foundation Books: New Delhi, 2010.

Karapinar, Baris and Christian Häberli (eds), *Food Crises and the WTO*, Cambridge University Press: Cambridge, 2010.

Kelman, Ilan, *Disaster Diplomacy: How Disaster Affect Peace and Conflict*, Routledge: London, 2012.

Khagram, Sanjeev, *Dams and Development: Transnational Struggles for Water and Power*, Oxford University Press: New Delhi, 2004.

Kiekby, et al., *The Earthscan Reader in Sustainable Development*, Earthscan: London, 1995.

Kutting, Gabriela (ed.), *Global Environmental Politics: Concepts, Theories and Case Studies*, Routledge: London, 2011.

Kutting, Gabriela, *Environment, Society and International Relations: Towards More Effective International Environmental Agreements*, Routledge: London, 2000.

Lewis, Devid, *Bangladesh: Politics, Economy and Civil Society*, Cambridge University Press: Cambridge, 2011.

Lipschutz, Ronnie D. (ed.), *On Security*, Columbia University Press: New York, 1998.

Mallikarjina, Raju, *Human Population and the Environment*, Akhand Publishing House: New Delhi, 2012.

Malone, Elizabeth L., *Debating Climate Change: Pathways through Argument to Agreement*, Earthscan: London, 2009.

Mathew, Richard A., Jon Barnett, et al., *Global Environmental Change and Human Security*, The MIT Press: Cambridge, MA, 2010.

Mathur, Jyotirmay, H. J. Wagner and N. K. Bansal (eds.), *Energy Security, Climate Change and Sustainable Development*, Anamaya Publisher: New Delhi, 2007.

Mazo, Jeffrey, *Climate Conflict: How Global Warming Threatens Security and What to Do about It*, The International Institute of Strategic Studies: London, 2010.

Miller, Marian A. L., *The Third World in Global Environmental Politics*, Open University Press: Buckingham, 1995.

Mishra, Omprakash (ed.), *Forced Migration in the South Asian Region: Displacement, Human Rights and Conflict Resolution*, Center for Refugee Studies, Jadavpur University: Kolkata, 2004.

Msangi, J. P. (ed.), *Combating Water Scarcity in Southern Africa: Case Studies from Namibia*, Springer: Dordrecht, 2014.

Oberthur, Sebastian and Hermann E. Ott, *The Kyoto Protocol: International Climate Policy for the 21st Century*, Springer: Berlin, 1999.

O'Neill, Brian C., F. Landis Mackellar and Wolfgang Lutz (eds.), *Population and Climate Change*, Cambridge University Press: Cambridge, 2001.

Pachauri, R.K. (ed.), *Dealing with Climate Change: Setting a Global Agenda for Mitigation and Adaptation*, The Energy and Resource Institute: New Delhi, 2010.

Paterson, Mathew, *Understanding Global Environmental Politics: Domination, Accumulation, Resistance*, Palgrave: Hampshire, 2000.

Pearson, Charles S., *Economics and the Challenge of Global Warming*, Cambridge University Press: London, 2011.

Pennington, Karrie Lynn and Thomas V. Cech, *Introduction to Water Resources and Environment*, Cambridge University Press: New York, 2010.

Petesch, Patti L., *North-South Environmental Strategies, Costs, and Bargains*, Overseas Development Council: Washington, DC, 1992.

Rajan Mukund Govind, *Global Environmental Politics: India and the North South Politics of Global Environmental Issues*, Oxford University Press: Oxford, 1997.

Roberts, Jane (ed.), *Environmental Policy: Critical Concepts in the Environment*, Vol. II, Routledge: London, 2007.

Roberts, Jane (ed.), *Environmental Policy: Critical Concepts in the Environment*, Vol. IV, Routledge: London, 2007.

Sachs, Wolfgang, *Global Ecology: A New Arena of Political Conflict*, Zed Books: London, 1993.

Sarah, Boulter, Jean Palutikof, David John Karoly and Daniela Guitart (eds.), *Natural Disasters and Adaptation to Climate Change*, Cambridge University Press: Cambridge, 2013.

Shiva, Vandana, *Staying Alive: Women, Ecology and Survival in India*, Zed Books: London, 1988.

Shiva, Vandana, *Stolen Harvest: The Hijacking of Global Food Supply*, South End Press: Canada, 2000.

Shiva, Vandana, *Water Wars: Privatization, Pollution and Profit*, South End Press: Cambridge, MA, 2002.

Shiva Vandana, et al., *Ecology and Politics of Survival: Conflicts Over Natural Resources in India*, SAGE: New Delhi, 1991.

Shukla, R., Subodh Kumar Sharma, Amit Garg, Sumana Bhattacharya and N. H. Rabindranath (eds.), *Climate Change and India: Vulnerability Assessment and Adaptation*, University Press: Hyderabad, 2003.

Sikka, Dr. Pawan, *Global Warming: India's Response to Climate Change, Disaster Mitigation and Adaptation*, Uppal Publishing House: New Delhi, 2010.

Singh Sanjay Kumar, *Environment Law and Climate Change*, SBS Publishers: New Delhi, 2010.

Solomon, S., D. Qin, M. Manning, Z. Chen, M. Marquis, K. B. Averyt, M. Tignor and H. L. Miller (eds.), *Climate Change: The Physical Science Basis: Contribution of Working Group I to the Fourth Assessment Report of the Intergovernmental Panel on Climate Change*, Cambridge University Press: Cambridge, 2007.

Sovacool, Benjamin K. (ed.), *The Routledge Handbook of Energy Security*, Routledge: London.

The Times of India, *Climate Change, Society and Sustainable Development*, Times Group Books: New Delhi, 2010.

Tomain, Joseph P., *Ending Dirty Energy Policy: Prelude to Climate Change*, Cambridge University Press: Cambridge, 2011.

Treidel, Holger, Jose Luis, Martin-Bordes and Jason J. Gurdak (eds.), *Climate Change Effects on Groundwater Resources: A Global Synthesis of Findings and Recommendations*, Taylor & Francis Group: London, 2012.

Tuathail, Gearóid Ó., Simon Dalby and Paul Routledge (eds.), *Geopolitics Reader*, Routledge: London, 1998.

Weatherbel, Donald E., *International Relations in Southeast Asia: The Struggle for Autonomy*, Rownam and Littlefield Publisher: Lanham, MD, 2009.

Wilson, G. A. and M. Junnti, *Unraveling Desertification: Policies and Actor Networks in Southern Europe*, Wageningen Academic Publisher: Wageningen, 2005.

Wollenberg, Eva and Marja-Liisa Tapio-Bistrom (eds.), *Climate Change Mitigation and Agriculture*, Earthscan: London, 2012.

Journal Articles in Print

Abahussain, Asma Ali, Anwar Sh. Abdu, Waleed K. Al-Zubari, Nabil Alaa El-Deen and Mahmmod Abdul-Raheem, "Desertification in the Arab Region: Analysis of Current Status and Trends", *Journal of Arid Environments*, Vol. 15, 2002, pp.521–545.

Ahmed, Imtiaz, "Teesta, Tipaimukh and River Linking: Danger to Bangladesh-India Relations", *Economic and Political Weekly*, Vol. XLVII, No. 16, April 21, 2012, pp.51–53.

Alam, Sarfaraz, "Environmentally Induced Migration from Bangladesh to India", *Strategic Analysis*, Vol. 27, No. 3, July–September 2003, pp.422–434.

Allan, Jen Iris and Peter Dauvergne, "The Global South in Environmental Negotiations: The Politics of Coalitions in REDD+", *Third World Quarterly*, Vol. 34, No. 8, 2013, pp.1307–1322.

Allenby, Braden R., Environmental Security: Concept and Implementation, *International Political Science Review*, Vol. 21, No. 1, 2000, pp.5–21.

Anderson, Emily K. and Hisham Zerriffi, "Seeing the Trees for the Carbon: Agroforestry for Development and Carbon Mitigation", *Climate Change*, Vol. 115, No. 3–4, December 2012, pp.741–757.

Asher, Manshi, "Renuka Dam: The Saga Continues", *Economic and Political Weekly*, Vol. XLVII, No. 32, August 11, 2012, pp.31–32.

Atteridge, Aaron, Manish Kumar Shrivastava and Himani Upadhyay, "Climate Policy in India: What Shapes International, National and State Policy?", *Ambio*, February, 2012, pp.68–77.

Balachandra, P., Darshini Ravindranath and N. H. Ravindranath, "Energy Efficiency in India: Assessing the Policy Regimes and Their Impacts", *Energy Policy*, Vol. 38, No. 11, November 2010, pp.6428–6438.

Bidwai, Praful, "Durban: Road to Nowhere", *Economic and Political Weekly*, Vol. XLVI, No. 53, December 31, 2011, pp.10–12.

Bose, Indrajit, Arnab Pratim Dutta and Souparno Banerjee, "Frozen at Gateway", *Down to Earth*, December 16–31, 2012, pp.32–46.

Bouchard, Christian and William Crumplin, "Climate Change, Sea Level Rise and Sustainable Development in Small Island States and Territories", *Journal of Indian Ocean Studies*, Vol. 19, No. 1, April 2011, pp.30–45.

Brack, Duncan and Katharina Umpfenbach, "Deforestation and Climate Change", *The World Today*, October 2009, pp.7–13.

Brandta, Urs Steiner and Gert Tinggaard Svendsen, "Hot Air in Kyoto, Cold Air in The Hague – the Failure of Global Climate Negotiations", *Energy Policy*, No. 30, 2002, pp.1191–1199.

Chakravarthi, Raghavan, 'South Says Equity the Touch-stone of Climate Proposals', *Third World Resurgence*, No. 13, September 1991, pp.5–6.

Church, John A., Neil J. White, Thorkild Aarup, W. Stanley Wilson, Philip L. Woodworth, Catia M. Domingues, John R. Hunter and Kurt Lambeck, "Understanding Global Sea Levels: Past, Present and Future", *Sustainability Science*, Vol. 3, No. 1, 2008, pp.9–22.

"COP16 Cancun, 2010: In Which Poor Countries Gave In", *Down to Earth*, January 1–15, 2011, pp.34–42.

Coper, Richard N., "Toward a Real Global Warming Treaty", *Foreign Affairs*, Vol. 72, No. 2, 1998, pp.66–79.

Corry, Olaf, "Securitization and 'Riskification': Second-order Security and the Politics of Climate Change", *Millennium*, Vol. 40, No. 2, 2012, pp.235–258.

Cousins, Stephanie, "UN Security Council: Playing a Role in the International Climate Change Regime?", *Global Change Peace and Security*, Vol. 25, No. 2, pp.191–210.

Dadwal, Shebonti Ray, "India's Renewable Energy Challenge", *Strategic Analysis*, Vol. 34, No.1, January 2010, pp.1–3.

Dalby, Simon, "Climate Change: New Dimensions of Environmental Security", *RUSI Journal*, Vol. 158, No. 3, 2013, pp.34–43.

Dalby, Simon, "Security, Modernity, Ecology: The Dilemmas of Post-Cold War security Discourse", *Alternatives: Global, Local Political*, Vol. 17, No. 1, Winter 1992, pp.95–134.

Das, Smriti, "The Strange Valuation of Forests in India", *Economic and Political Weekly*, Vol. XIV, No. 9, February 27, 2010, pp.16–18.

Dixon, Thomas Homer, "Environmental Scarcities, State Capacity and Civil Violence", *Bulletin of the American Academy of Arts and Sciences*, Vol. 48, No. 7, April 1995, pp.26–33.

Dubash, Navroz K. and Ann Florini, "Mapping Global Energy Governance", *Global Policy*, Vol. 2, September 2011, pp.6–18.

Dutt, Gautam, "A Climate Agreement Beyond 2012", *Economic Political Weekly*, Vol. XLIV, No. 45, November 7, 2009, pp.39–49.

Dutt, Gautam and Fabian Gaioli, "Coping with Climate Change", *Economic and Political Weekly*, Vol. 42, No. 42, October 20–26, 2007, pp.4239–4250.

Farber, Daniel A., "Basic Compensation for Victims of Climate Change", *University of Pennsylvania Law Review*, Vol. 155, No. 6, 2007, pp.1605–1656.

Feibiao, Xu and Jiang Li, "Food Security in the Context of Global Climate Change", *Contemporary International Relations*, Vol. 21, No. 3, May–June, 2011, pp.23–33.

"For Land and Water", *Economic and Political Weekly*, Vol. XLVI, No. 34, August 20, 2011, pp.9–10.

Froggattand, Antony and Michael A. Levi, "Climate and Energy Security Policies and Measures: Synergies and Conflicts", *International Affairs*, Vol. 85, No. 6, 2009, pp.1129–1141.

Gadgil, Madhav and Ramachandra Guha, "Ecological Conflicts and the Environmental Movement in India", *Development and Change*, Vol. 25, 1994, pp.101–136.

Gautam, P.K., "Climate Change and Conflict in South Asia", *Strategic Analysis*, Vol. 36, No. 1, January 2012, pp.32–40.

Geevan, C.P., "National Environment Policy: Ascendance of Economic Factors", *Economic and Political Weekly*, October 23, 2004.

Ghoble, Vrushal T., "The Economics of Natural Gas: Its Geopolitical Implications", *World Affairs*, Vol. 17, No. 2, Summer 2013, pp.108–123.

Girjesh Pant, "The Future of Energy Security", *South Asian Survey*, Vol. 17, No. 1, 2010, pp.31–43.

Gleick, Peter H., "Climate Change and International Politics: Problem Facing Developing Countries", *Ambio*, Vol. 18, No. 6, 1989, pp.333–339.

Godement, Francois, "China's Energy Policy: From Self-sufficiency to Energy Efficiency", *The International Spectator*, Vol. 42, No. 3, September 2007, pp.391–397.

Grainger, Alan, Mark Stafford Smith, Victor R. Squires and Edward P. Glenn, "Desertification and Climate Change: The Case for Greater Convergence", *Mitigation and Adaptation Strategies for Global Change*, Vol. 5, 2000, pp.361–377.

Granahan, M.C., G.D. Balk and B. Anderson, "The Rising Tide: Assessing the Risks of Climate Change and Human Settlements in Low Elevation Coastal Zones", *Environment and Urbanization*, Vol. 19, No. 1, 2007, pp.17–27.

Gupta, Joyeeta, "India and Climate Change Policy: Between Diplomatic Defensiveness and Industrial Transformation", *Energy and Environment*, Vol. 12, Nos. 2 and 3, 2001, pp.217–236.

Gupta, Shreekant, "Dithering on Climate Change", *Economic and Political Weekly*, Vol. 37, No. 51, December 21–27, 2002, pp.5073–5076.

Hameiri, Shahar and Lee Jones, "The Politics and Governance of Non Traditional Security", *International Studies Quarterly*, Vol. 57, No. 3, September 2013, pp.462–473.

"Helping Forests Disappear", *Economic and Political Weekly*, Vol. XLVII, No. 8, February 23, 2013, p.8.

Holtermann, Ole Magnus, Theisen Helge and Halvard Buhaug, "Climate Wars? Assessing the Claim That Drought Breeds Conflict", *International Security*, Vol. 36, No. 3, Winter 2011/12, pp.79–106.

Hook, Mikael and Xu Tang, "Depletion of Fossil Fuels and Anthropogenic Climate Change – A Review", *Energy Policy*, Vol. 52, January 2013, pp.797–809.

Hurrell, Andrew and Sandeep Sengupta, "Emerging Powers, North-South Relations and Global Climate Politics", *International Affairs*, Vol. 88, No. 3, 2012, pp.463–484.

Iyer, Ramaswamy R., "Governance of Water: The Legal Questions," *South Asian Survey*, Vol.17, No. 1, 2010, pp.147–157.

Jacoby, Henry D., Prinn G. Ronald and Richard Schmalensee, "Kyoto's Unfinished Business", *Foreign Affairs*, Vol. 77, No. 4, July–August 1998, pp.58–65.

Jayaram, Dhanasree, "India-China Relations: The Way Forward Through Renewable Energy Diplomacy", *Defence and Diplomacy*, Vol. 2, No. 2, January–March 2013, pp.91–100.

Jayaraman, T., Tejal Kanitkar and Mario D'souza, "Deconstructing the Climate Blame Game", *Economic and Political Weekly*, Vol. XLV, No. 1, January 2, 2010, pp.13–15.

Joshi, Vijay, and Urjit R. Patel, "India and a Carbon Deal", Economic and Political Weekly, Vol. XLIV, No. 31, August 1, 2009, pp.71–77.

Kelman, Ilan and Jennifer J. Wes, "Climate Change and Small Island Developing States: A Critical Review", *Ecological and Environmental Anthropology*, Vol. 5, No. 1, 2009, pp.1–16.

Keohane, Robert O. and David G. Victor, "The Regime Complex for Climate Change", *Perspectives on Politics*, Vol. 9, No. 1, March 2011, pp.7–23.

Keohane, Robert O. and David G. Victor, "The Transnational Politics of Energy", *Daedalus*, Vol. 142, No. 1, Winter 2013, pp.97–109.

Khan, Mizan R., "From Cancun to Durban: Is There Any Likelihood of a New Climate Regime?", *BIIS Journal*, Vol. 32, No. 1, January 2011, pp.41–63.

Khor, Martin, "Complex Implications of the Cancun Climate Conference", *Economic and Political Weekly*, Vol. XLV, No. 52, December 25, 2010, pp.10–15.

Khor, Martin, "Doha 2012: A Climate Conference of Low Ambition", *Economic and Political Weekly*, Vol. XLVIII, No. 3, 2013, pp.18–21.

Kiran Shrestha, Reshmi, Rhodante Ahlers, Marloes Bakker and Joyeeta Gupta, "Institutional Dysfunction and Challenges in Flood Control: A Case Study of the Kosi Flood 2008", *The Economic and Political Weekly*, Vol. XLV, No. 2, January 9, 2010, pp.45–53.

Kumar, K. S. Kavi, Priya Shyamsundar and A. Arivudai Nambi, "Economics of Climate Change Adaptation in India", *The Economic and Political Weekly*, Vol. XLV, No. 18, May 1, 2010, pp.25–29.

Kumar, Vinod and A. K. Chopra, "Impact of Climate Change on Biodiversity of India with Special Reference to Himalayan Region: An Overview", *Journal of Applied and Natural Science*, Vol. 1, No. 1, 2009, pp.117–122.

Latif, Katharina, "New Energy for Investors", D+C *(Development and Cooperation)*, Vol. 39, July/August 2012, p.283.

Lee, Bernice, "Managing the Interlocking Climate and Resource Challenges", *International Affairs*, Vol. 85, No. 6, 2009, pp.1101–1116.

Lele, Sharachchandra, "Rethinking Sustainable Development", *Current History*, Vol. 112, No. 757, November 2013, pp.311–316.

Levi, Michael A., "Copenhagen's Inconvenient Truth: How to Salvage the Climate Conference", *Foreign Affairs*, Vol. 88, No. 5, September–October 2009, pp.92–104.

Loftus, Alex, "Rethinking Political Ecologies of Water", *Third World Quarterly*, Vol. 30, No. 5, 2009, pp.953–968.

Martin, Adrian, "Environmental Conflict between Refugee and Host Communities", *Journal of Peace Research*, Vol. 42, No. 3, May 2005, pp.329–346.

Martin, Susan, "Climate Change, Migration and Governance", *Global Governance*, Vol. 16, No. 9, January–March 2010, pp.397–414.

Mathews, Jessica Tuchman, "Redefining Security", *Foreign Affairs*, Vol. 68, No. 2, Spring 1989, pp.162–177.

May, Bernhard, "Energy Security and Climate Change", *South Asian Survey*, Vol. 17, No. 1, 2010, pp.19–30.

Mayer, Maximilian, "Chaotic Climate Change and Security", *International Political Sociology*, Vol. 6, No. 2, June 2012, pp.165–185.

McCarl, Bruce A., Mario A. Fernandez, Jason P. H. Jones and Marta Wlodarz, "Climate Change and Food Security", *Current History*, Vol. 112, No. 750, January 2013, pp.33–37.

Methmann, Chris and Delf Rothe, "Politics for the Day After Tomorrow: The Logic of Apocalypse in Global Climate Politics", *Security Dialogue*, Vol. 43, No. 4, 2012, pp.323–344.

Michael, Grubb, "Kyoto and the Future of International Climate Change Responses: From Here to Where", *International Review for Environmental Strategies*, Vol. 5, No. 1, 2004, pp.15–38.

Moniz, Maria Da Graca Canto, "India's Carbon Governance: The Clean Development Mechanism", *Future of Food: Journal on Food, Agriculture and Society*, Vol. 1, No. 1, Summer 2013, pp.5–15.

Moutinho, P., M. Santilli, S. Schwartzman and L. Rodrigues, "Why Ignore Tropical Deforestation? A Proposal for Including Forest Conservation in the Kyoto Protocol", *Unasylva 222*, Vol. 56, 2005, pp.27–30.

Myers, Norman, "Environment and Security", *Foreign Policy*, No. 74, Spring 1989, pp.23–41.

Myers, Norman, "Environmental Refugees: A Growing Phenomenon of the 21st Century", *Philosophical Transactions of the Royal Society of London Series, Biological Sciences*, Vol. 357, No. 1420, 2002, pp.609–613.

Myers, Norman, "Environmental Refugees in a Globally Warmed World", *Bio Science*, Vol. 43, No. 11, December 1993, pp.752–761.

Nicholl, Robert J., "Planning for the Impacts of Sea Level Rise", *Oceanography*, Vol. 24, No. 2, 2011, pp.144–157.

Ott, Hermann E., "Climate Change: An Important Foreign Policy Issue", *International Affairs*, Vol. 77, No. 2, 2001, pp.277–296.

Padukone, Neil, "Climate Change in India: Forgotten Threats, Forgotten Opportunities", *Economic and Political Weekly*, Vol. XLV, No. 22, May 29, 2010, pp.47–54.

Panda, Architesh, "Climate Refugees: Implications for India", *Economic and Political Weekly*, Vol. XLV, No. 20, May 15, 2010, pp.76–79.

Pandit, M. K., Navjot S. Sodhi, Lian Pinkoh, Arun Bhaskar and Barry W. Brook, "Unreported Yet Massive Deforestation Driving Loss of Endemic Biodiversity In Indian Himalaya", *Biodiversity Conservation*, Vol. 16, 2007, pp.153–163.

Pant, Girjesh, "The Future of Energy Security Through a Global Restructuring", *South Asian Survey*, Vol. 17, No. 1, 2010, pp.31–43.

Rabindranath, N. H., "The Copenhagen Accord", *Current Science*, Vol. 98, No. 6, March 25, 2010.

Raghunandan, D., "Durban Platform: Kyoto Negotiations Redux", *Economic and Political Weekly*, Vol. XLVI, No. 53, December 31, 2011.

Raghunandan, D., "Kyoto Is Dead, Long Live Durban?", *Economic and Political Weekly*, Vol. XLV, No. 52, December 25, 2010, pp.16–20.

Rajamani, Lavanya, "The Changing Fortunes of Differential Treatment in the Evolution of International Environmental Law", *International Affairs*, Vol. 88, No. 3, 2012, pp.615–620.

Ravindranath, N. H., N. V. Joshi, R. Sukumar and A. Saxena, "Impact of Climate Change on Forests in India", *Current Science*, Vol. 90, No. 3, February 10, 2006, pp.354–361.

Ravindranath, N. H., Rajiv Kumar Chaturvedi and Indu K. Murthy, "Forest Conservation, Afforestation and Reforestation in India: Implications for Forest Carbon Stocks", *Current Science*, Vol. 95, No. 2, July 25, 2008, pp.216–222.

Reuveny, Rafael, "Climate Change-Induced Migration and Violent Conflict", *Political Geography*, Vol. 26, 2007, pp.656–673.

Ruhl, Christof, "Global Energy after the Crisis," *Foreign Affairs*, Vol. 89, No. 2, March/April 2008–09, pp.63–75.

Sabur, A. K.M. Abdus, "Disaster Management System in Bangladesh: An Overview", *India Quarterly*, Vol. 68, No. 1, March 2012, pp.29–47.

Salehyan, Idean, "From Climate change to Conflict? No Consensus Yet", *Journal of Peace Research*, Vol. 45, No. 3, May 2008, pp.315–326.

Sanyal, Mukul, "Global Vision for Rio+20 and Beyond", *Economic and Political Weekly*, Vol. XLVI, No. 40, October 1, 2011, pp.25–30.

Scheffran, J. and A. Battaglini, "Climate and Conflicts: The Security Risks of Global Warming", *Regional Environmental Change*, Vol. 11, No. 1, March 2011, pp.S27–S39.

Scott, Shirley V., "Climate Change and Peak Oil as Threats to International Peace and Security: Is It Time for Security Council to Legislate", *Melbourne Journal of International Law*, Vol. 9, 2008, pp.1–3.

Sharan, Shyam, "Climate Change: The Road to Cancun", *Indian Foreign Affairs Journal*, Vol. 5, No.1, January–March 2010, pp.1–5.

Siddiqi, Toufiq, "China and India: More Cooperation Than Competition in Energy and Climate Change", *Journal of International Affairs*, Vol. 64, No. 1, Fall/Winter 2010, pp.73–90.

Singh, Bhupendra Kumar, "India's Energy Security: Challenges and Opportunities", *Strategic Analysis*, Vol. 34, No. 6, November 2010, pp.799–805.

Singh, Nigel, "The Role of the National Solar Mission in Climate Change Mitigation and the Twin Objective of Energy Security", *Strategic Analysis*, Vol. 36, No. 2, March 2012, pp.260–275.

Singh, R. K. Ranjan, "Tipaimukh Is a Death Trap for Indigenous People", *The Ecologist Asia*, Vol. 11, No. 1, January–March 2003, pp.76–79.

Sinha, Uttam Kumar, "Climate Summit at Copenhagen: Negotiating the Intractable", *Strategic Analysis*, Vol. 33, No. 6, November 2009, pp.795–799.

Smith, Heather A., "Facing Environmental Security", *Journal of Military and Strategic Studies*, Winter 2000 – Spring 2001, pp.36–48.

Stevenson, Hayley, "India and International Norms of Climate Governance: A Constructivist Analysis of Normative Congruence", *Review of International Studies*, Vol. 37, 2011, pp.997–1019.

Streck, Charlotte, "New Partnerships in Global Environmental Policy: The Clean Development Mechanism", *Journal of Environment & Development*, Vol. 13, No. 3, September 2004, pp.295–322.

Thwaites, Joe, "Pakistan's Shifting Discourse on Climate Change at the United Nations Security Council", *Journal of International Service*, Vol. 3, No. 1, Spring 2014, pp.67–86.

Tripathi, Narendra Kumar, "Hydropower in Asia: Spinning a Dependence and Interdependence Binary", *South Asian Survey*, Vol. 17, No. 2, September 2010, pp.219–235.

Urdal, Henrik, "People Vs. Malthus: Population Pressure, Environmental Degradation, and Armed Conflict Revisited", *Journal of Peace Research*, Vol. 42, No. 4, July 2005, pp.417–434.

Vyas, V. S. and V. Ratna Reddy, "Assessment of Environmental Policies and Policy Implementation in India", *Economic and Political Weekly*, January 10, 1998, pp.48–54.

Wangkheirakpam, Ramananda, "Lessons from Loktak", *The Ecologist Asia*, Vol. 11, No. 1, January–March 2003, pp.19–24.

William, Angella, "Turning the Tide: Recognising Climate Change Refugees in International Law", *Law and Policy*, Vol. 30, No. 4, October 2008, pp.502–529.

Wirth, David A., "Climate Chaos", *Foreign Policy*, No. 74, Spring 1989, pp.3–22.

Wu, Fuzuo, "Sino-Indian Climate Cooperation: Implications for the International Climate Change Regime", *Journal of Contemporary China*, Vol. 21, No. 77, September 2012, pp.827–843.

Wyman, Katrina Miriam, "Response to Climate Migration", *Harvard Environmental Law Review*, Vol. 37, 2013, pp.168–216.

Yumkella, Kandekh K., "Sustainable Energy for All: Towards Rio+20", *UN Chronicle*, No. 1 and 2, 2012, pp.18–21.

Newspaper Articles in Print

"A Poor Relief Effort", *The Hindu*, December 29, 2004.

Anand, Annu, "The Perils of Progress", *The Hindu*, March 22, 2013.

Anuradha, R. V., "Trade Law: The Road from Copenhagen", *The Times of India*, December 17, 2009.

"Asia Faces Risks of Disease, Hunger: Document-Working Group-II Contribution to the IPCC 4th Assessment Report", *The Asian Age*, April 7, 2007.

"Averting Water Crisis", *The Hindu*, July 9, 2007.

Balaji, S., "Biodiversity Challenges Ahead", *The Hindu*, May 27, 2010.

Bodh, Anand, "In HP, Uttarakhand Waiting to Happen", *The Times of India*, July 3, 2013.

Chader, Parkash, "How to Talk Climate Change in Paris", *The Hindu*, December 17, 2014.

Chakraborti, Suman and Rohit Khanna, "World Bank May Take Back Coastal Funds", *The Times of India*, June 11, 2013.

Chaturvedi, Satyarat, "GM Crops Are No Way Forward", *The Hindu*, August 24, 2012.

Chellaney, Brahma, "Ensuring Resource Security: From a Local Problem to a Global Challenge", *The Times of India*, March 30, 2010.

Dhar, Aarati, "A Man-Made Disaster, Say Environmentalists", *The Hindu*, June 21, 2013.

"Diabolic Game", *Deccan Herald*, August 13, 2012.

"End Biopiracy", *Deccan Herald*, October 10, 2012.

"Fresh Flood Threat to Krishna, Guntur", *The Hindu*, October 6, 2009.

George, Varghese K. and Chetan Chouhan, "The Law Catches Up with Vedanta", *The Hindustan Times*, August 25, 2010.

"GM Crops Not For India: Parliamentary Panel", *The Economic Times*, August 10, 2012.

Gray, Louise, "Cancun Climate Change Summit: Small Island States in Danger of 'Extinction' ", *The Telegraph*, December 1, 2010.

Haq, Zia and Gautam Choudhury, "India's Mid-Summer Nightmare", *The Hindustan Times*, July 29, 2012.

Hegde, Pandurang, "Need to Preserve Our Bio Diversity", *Deccan Herald*, October 10, 2012.

"How Snow Melt Fed Swollen Rivers", *Times of India*, July 3, 2013.

"India Leads in CDM Projects", *The Hindu*, 18 February 2005.

"India Stares at Uneven Monsoons", *The Times of India*, June 20, 2013.

Iyer, Ramaswamy R., "For Rejuveneting, Not Reengineering, the Ganga", *The Hindu*, July 16, 2014.

Jacobs, Michael, "Lima Deal Represents a Fundamental Change in Global Climate Regime", *The Guardian*, December, 15, 2014.

Jayaraman, T., "India and Climate Talks Imperatives", *The Hindu*, November 18, 2013.

Jayaraman, T., "Post-Durban, India Has Its Task Cut Out", *The Hindu*, December 22, 2011.

"Jayram Defends Nuancing India's Position at Cancun", *The Hindu*, December 25, 2010.

Kasturi, Charu Sudan, "Climate Change Induced Food Nightmare Stares India", *The Hindustan Times*, June 7, 2011.

Khasnobis, H., "Environmental Refugees: Climate Change, Conflict and Forced Migration", *The Statesman*, April 6, 2012.

Krishnan, Ananth, "China Reassures India on Dam Projects", *The Hindu*, November 6, 2009.

"Laila: 6 Andhra Districts Hit, 50,000 Evacuated", *Indian Express*, May 21, 2010.

Mandal, Caesar, "Exodus from Tide Country to Kolkata", *The Times of India*, June 1, 2009.

Mehdudia, Sujay, "Bartering Green Ideas", *The Hindu*, April 25, 2013.

Menon, Meena, "Climate Change Summit: Countries Demand Focus on Adaptation and Finance", *The Hindu*, December 5, 2014.

Menon, Meena, "Doha Dithers on Equity", *The Hindu*, December 15, 2012.

Menon, Meena, "Green Energy and Beyond", *The Hindu*, April 20, 2013.

Menon, Meena, "India to Press for Green Fund Operationalisation at Lima Climate Meet", *The Hindu*, November 30, 2014.

Menon, Meena, "Lima: A New Low for Climate Action", *The Hindu*, December 22, 2014.

Nandi, Jayashree, "Depleting Forest Cover Could Have Grave Impact", *The Times of India*, June 21, 2011.

Nandi, Jayashree, "World Environment Day 2013: Remembering Silent Valley Movement", *The Times of India*, June 5, 2013.

"New Coastal Management Rules", *The Hindu*, June 2, 2008.

Pandit, Maharaj K., "Nature Avenges Its Exploitation", *The Hindu*, June 21, 2013.

Parsai, Gargi, "India Can Go Ahead with Kishenganga", *The Hindu*, February 19, 2013.

Parsai, Gargi, "Water Policy Draft Favours Privatisation of Services", *The Hindu*, January 21, 2012.

Pinjarkar, Vijay, "New Working Plan Code Gives New Approach to Forestry", *The Times of India*, May 14, 2014.

"PM Calls as Assam Floods Worsen", *The Times of India*, July 9, 2013.

PTI, "Include More Adaptation Efforts in Paris Climate Deal: India", *The Hindu*, December 4, 2014.

PTI, "Lima Climate Deal Addresses Our Concern: India", *The Hindustan Times*, December 14, 2014.

PTI, "UN Climate Summit Begins in Peru", *The Hindu*, December 20, 2014.

Qureshy, Ahtesham, "Rio Summit Not Satisfying for Developing Nations", *The Hindustan Times*, June 16, 1992.

Raj, N. Gopal, "Melting Glaciers, More Rain to Swell Himalayan Rivers", *The Hindu*, June 2, 2014.

Rajamani, Lavanya, "Deconstructing Durban", *The Indian Express*, December 15, 2011.

Ramachandran, R., "Climate Change and the Indian Stand", *The Hindu*, July 28, 2009.

Raman, T. R. Shankar, "One Earth One Chance", *The Hindu*, June 4, 2011.

"Rivers in Danger", *The Hindu*, March 24, 2007.

Saxena, Ketaki, "Copenhagen Shame: Trust Deficit Galore", *The Pioneer*, December 25, 2009.

Sethi, Nitin, "G77+China Group Walk Out of Loss and Damage Talks", *The Hindu*, November 20, 2013.

Shashi Tharoor, "The Politics of Energy", *The Asian Age*, October 12, 2012.

Siddharth, Gautam, "Restoring the Himalayas", *The Times of India*, July 4, 2013.

Singh, Mahim Pratap, "Fertilizer Subsidy Hurts Everyone", *The Hindu*, July 10, 2014.

Singh, Mahim Pratap, "Rajasthan Tops in Solar Energy", *The Hindu*, May 10, 2013.

Thakur, Atul, "Floods Affect 30 Million Indians Every Year", *The Times of India*, June 20, 2013.

"The Roadmap to Mexico – Climate Convention to Forward", *The Times of India*, February 4, 2010.

"UN Meet on Biodiversity", *The Statesman*, October 9, 2012.

"Using Groundwater Wisely", *The Hindu*, October 9, 2007.

"Vedanta Project Fails Environment Test", *The Statesman*, August 25, 2010.

Venkatesan, J., "Fear of Nuclear Disaster Has No Basis", *The Hindu*, May 7, 2013.

"Water Sector Needs Purposeful Action to Match India's Long Term Needs", *The Economic Times*, March 29, 2014.

Web Articles and Papers

Ahluwalia, Montek S., and Patel, Utkarsh, "The Glasgow summit on climate change: What has it achieved?", *Live mint*, November 14, 2021. Available at The Glasgow summit on climate change: What has it achieved? (livemint.com) Accessed on January 18, 2022.

Al Arabiya News, *Two Dead and 400 Injured in Algeria Riots*, January 8, 2011. Available at http://www.alarabiya.net/articles/2011/01/08/132669.html Accessed on December 14, 2014.

Angre, Ketki, "Rio + 20-The Indian Perspective", *NDTV*, June 23, 2012. Available at www.ndtv.com Accessed on April 24, 2014.

"Assam Flood Toll Rises to 121", *The Hindu*, July 7, 2012. Available at www.thehindubusinessline.com/.../national-mission-on-clean-coal-techno... Accessed on August 14, 2014.

Barnett John, "Security in Asia: Issues and Implications for Australia", Melbourne-Asia Policy Papers No. 9, The University of Melbourne: Melbourne, March 2007. Available at http://www.greencrossaustralia.org/media/81235/asialink%20-%20 climate%20change%20and%20security%20in%20asia%20-%20issues%20 and%20implications%20for%20australia.pdf Accessed on December 21, 2014.

Bassetti, Francesco, "*Environmental Migrants: Up to 1 Billion by 2050*, May 22, 2019. Available at climateforesight.eu Accessed on January 22, 2022.

Basu, Jayanta, "Climate Emergency CoP 25: Loss and Damage 'Fighting Out' in Madrid", *Down to Earth*, December 13, 2019. Available at downtoearth.org.in Accessed on January 23, 2022.

Benninghoff, Virginia, "Prioritizing Fossil-Fuel Subsidy Reform in the UNFCCC Process: Banerjee Srestha, Green Tribunal Halts Construction on Puri Coast", *Down to Earth*, December 7, 2013. Available at http://www.downtoearth.org.in/ content/green-tribunal-halts-construction-puri-coast Accessed on July 31, 2014.

Bergesen, Helge Ole and Georg Parmann (eds.), *Green Globe Yearbook of International Co-operation on Environment and Development*, Oxford University Press: Oxford, 1994. Available at http://www.fni.no/ybiced/94_06_toulmin.pdf Accessed on November 15, 2014.

Betz, Joachim, "India's Turn in Climate Policy: Assessing the Interplay of Domestic and International Policy Change", GIGA Working Paper, No. 9, March 2012. Available at www.giga-hamburg.de/workingpapers Accessed on April 25, 2014.

Beyond Copenhagen, "Indian NGOs Bust the Myth of Climate Smart Agriculture", December 6, 2011. Available at http://beyondcph.blogspot.in/2011/12/indian-ngos-bust-myth-of-climate-smart.html Accessed on July 13, 2014.

Bhasin, Shikha, Tobias Engelmeier and Felix Schimdt, "After Cancún India's New Role as an International Deal Maker", *Friedrich Ebert Stifung, Perspective*, April 2011. Available at http://library.fes.de/pdf-files/iez/08145.pdf Accessed on March 21, 2015.

Bhattacharjya, Souvik and Nitya Nanda, *Potential Impact of Carbon Barriers to Trade: The Case of India's Exports to the US Under Border Tax Adjustment*, TERI: New Delhi, December 2012. Available at www.teriin.org/projects/nfa/pdf/ Working_paper3.pdf Accessed on August 13, 2014.

Bose, Indrajit, "Fishing in New Waters", *Down to Earth*, February 15, 2014. Available at http://www.downtoearth.org.in/content/fishing-new-waters Accessed on March 21, 2015.

Brger, Julian, "Darfur Conflict Heralds Era of Wars Triggered by Climate Change, UN Report Warns", *The Guardian*, June 23, 2007. Available at http://www. theguardian.com/environment/2007/jun/23/sudan.climatechange Accessed on September 20, 2014.

Carpenter, Chad, "Taking Stock of Durban: Review of Key Outcomes and the Road Ahead", *UNDP Environment and Energy Group*, April 2012. http://www. undpcc.org/docs/Bali%20Road%20Map/English/UNDP_Taking%20Stock%20 of%20Durban.pdf Accessed on June 7, 2014.

Carrington, Damian, "What Is 'Loss and Damage' and Why Is It Critical for Success at Cop26?", *The Guardian*, November 13, 2021. Available at https://www. theguardian.com/environment/2021/nov/13/what-is-loss-and-damage-and-why-is-it-critical-for-success-at-cop26 Accessed on January 23, 2022.

Church et al., "Understanding Global Sea Levels: Past, Present and Future", *Sustainability Science*, Vol. 3, No. 1, pp.9–22. Available at http://unesdoc.unesco.org/ images/0015/001561/156135E.pdf Accessed on January 3, 2015.

Conference Statement, *The Changing Atmosphere: Implications for Global Security*, Conference held in Toronto Canada, June 27–30, 1988. Available at http://www.cmos.ca/ChangingAtmosphere1988e.pdf Accessed on May 15, 2014.

Cullet, Philippe, "Water Law in India: Overview of Existing Framework and Proposed Reforms", *International Environmental Law Research Centre*, 2007. Available at http://www.ielrc.org/content/w0701.pdf Accessed on July 17, 2014.

"Cyclone Sidr in Bangladesh", *The Guardian*, November 19, 2007. Available at http://www.theguardian.comTheGuardian/flash/page/0,,2212365,00.html Accessed on December 15, 2014.

Danda, Anamitra Anurag, Gayathri Sriskanthan, Asish Ghosh, Jayanta Bandyopadhyay and Sugata Hazra, *Indian Sundarbans Delta: A Vision, World Wide Fund for Nature-India: New Delhi*, 2011. Available at http://www.indiaenvironmentportal.org.in/files/indian_sundarbans_deltaavision.pdf Accessed on July 28, 2013.

Dandekar, Parineeta and Himanshu Thakkar, *Ecological Management of Rivers in India: A Long Road Ahead*, January 2012. Available at sandrp.in/rivers/Ecological_Management_of_Rivers_in_India_Jan_2012... Accessed on July 19, 2014.

Dharmmala, Nikhilesh, "The Role of Carbon Capture and Storage in India's 'Hard to Abate' Industries", *The Wire*, February 19, 2021. Available at https://science.thewire.in/economy/energy/what-role-does-carbon-capture-and-storage-play-in-indias-climate-puzzle/ Accessed on January 24, 2022.

Diallo, Hama Arba, "Seizing the Chance", *Our Planet (The Magazine of the UNEP)*, Vol. 17, No. 1, pp.9–10. Available at http://www.ourplanet.com/imgversn/171/Hama%20Arba%20Diallo. pdf Accessed on November 15, 2014.

Dubrin, Adam and George Bowden, "COP26: UK Pledges £290m to Help Poorer Countries Cope with Climate Change", *BBC News*, November 8, 2021. Available at https://www.bbc.com/news/uk-59202129 Accessed on January 19, 2022.

D'Souza, John and Raajen Singh, "Missing the Community for the Woods: Forests, Communities and Climate change In India, A Climate Education Booklet for INECC (Indian Network for Ethics in Climate Change)", *Centre for Education & Documentation*, 2010. Available at www.ced.org.in/docs/inecc/forest booklet/2forest_booklet.pdf Accessed on December 29, 2014.

Earth Negotiations Bulletin, Vol. 8, No. 52, August 31, 2013. Available at http://www.iisd.ca/download/pdf/enb0852e.pdf Accessed on June 23, 2014.

Eckstein, David, Vera Künzel and Laura Schäfe, *Global Climate Risk Index, 2021: Who Suffers Most from Extreme Weather Events? Weather-Related Loss Events in 2019 and 2000–2019, Briefing Paper*, 2021. Available at https://germanwatch.org/sites/default/files/Global%20Climate%20Risk%20Index%202021_1.pdf Accessed on January 22, 2022.

Eriksson, Mats et al., *The Changing Himalayas: Impact of Climate Change on Water Resources and Livelihoods in the Greater Himalayas* (ICIMOD), December 11, 2009. Available at www.worldwatercouncil.org/...water.../climate_change/PersPap.01._The... Accessed on June 25, 2013.

Frangoul, Anmar, "China's Shock Climate Deal with the U.S. Sparks Some Cautious Optimism", *CNBC*, November 11, 2021. Available at COP26: U.S.-China declaration on climate welcomed (cnbc.com) Accessed on January 23, 2022.

Fricke, Laura, "Implementation of the Clean Development Mechanism in India", *UW Bothell Policy Journal*, Spring 2008, pp.63–69. Available at http://uwbpolicyjournal.files.wordpress.com/2012/06/implementation-of-the-clean-development.pdf Accessed on January 8, 2013.

Froggatt, Antony, "China, EU and US Cooperation on Climate and Energy, Research Paper", *Chatham House*, March 29, 2021. Available at https://www.chatham-house.org/2021/03/china-eu-and-us-cooperation-climate-and-energy Accessed on January 20, 2021.

Georgina, Rannard and Francesca Gillett, "COP26: World Leaders Promise to End Deforestation by 2030", *BBC News*, November 2, 2021. Available at https://www.bbc.com/news/science-environment-59088498 Accessed on January 20, 2022.

Ghosh, Padmaparna, "I Want to Position India as a Proactive Player: Jairam Ramesh", *Mint*, September 29, 2009. Available at http://www.livemint.com/Politics/h97ogi3qEaToXuP9T8YTuI/I-want-to-position-India-as-a-proactive-player-Jairam-Rames.h Accessed on January 17, 2014.

Global Energy Network Institute, *Overview of Sustainable Renewable Energy Potential of India*, January 2010. Available at http://www.geni.org/globalenergy/research/renewable-energy-potential-of-india/Renewab%20Energy%20Potential%20for%20India.pdf Accessed on September 8, 2013.

Global Water Partnership, *A Handbook for Integrated Water Resources Management in Basins*, 2009. Available at http://www.gwp.org/Global/ToolBox/References/A%20Handbook%20for%20Integrated%20Water%20Resources%20Management%20in%20Basins%20%28INBO,%20GWP%202009%29%20ENGLISH.pdf Accessed on May 28, 2014.

Greenpeace, *Ensuring Our Food Security*. Available at http://www.greenpeace.org/india/en/What-We-Do/Sustainable-Agriculture/ Accessed on July 13, 2014.

Greenpeace, *Reviving Our Soils*, February 15, 2011. http://www.greenpeace.org/india/en/What-We-Do/Sustainable-Agriculture/Fertiliser-campaign/ Accessed on July 13, 2014.

Griffiths, Tom, "Seeing 'RED'?: 'Avoided Deforestation' and the Rights of Indigenous Peoples and Local Communities", *Forest People's Programme*, June 2007. Available at http://www.forestpeoples.org/sites/fpp/files/publication/2010/01/avoideddeforestationredjun07eng_0.pdf Accessed on February 4, 2015.

Gupta, Rajiv K., *Climate Change Department, Government of Gujarat, Climate Change and Water Resources Management in Gujarat*. Available at www.aragon.es/estaticos/GobiernoAragon/.../Rajiv%20Gupta.pdf Accessed on July 28, 2013.

Hall, Nina, "Climate Change and Institutional Change in UNHCR", Conference Paper for UNU-EHS Summer Academy on Protecting Environmental Migration: Creating New Policy and Institutional Frameworks, July 25–31, 2010. Available at http://www.ehs.unu.edu/file/get/5404 Accessed on June 17, 2014.

Haque, Ubydul, Masahiro Hashizume, Korine N. Kolivras, Hans J. Overgaard, Bivash Das and Taro Yamamoto, "Reduced Death Rates from Cyclones in Bangladesh: What More Needs to Be Done?", *Bulletin of the World Health Organization*, October 24, 2011. Available at http://www.who.int/bulletin/volumes/90/2/11-088302/en/ Accessed on December 15, 2014.

Harrabin, Roger, "Indian Farm Makes Carbon Capture Breakthrough", *The Guardian*, January 4, 2017. Available at https://www.theguardian.com/environment/2017/jan/03/indian-firm-carbon-capture-breakthroughcarbonclean#:~:text=A%20breakthrough%20in%20the%20race,it%20to%20make%20baking%20soda Accessed on January 12, 2022.

Hawkins, William R., "China – India Accord to Scuttle UN Climate Treaty", *American Thinker*, October 23, 2009. Available at www.americanthinker.com/2009/10/chinaindia_accord_to_scuttle_u.html Accessed on March 19, 2014.

Hazra, Arnab Kumar, "History of Conflict over Forests in India: A Market Based Resolution", Working Paper Series, Julian L. Simon Centre for Policy Research, April 2002. Available at www.libertyindia.org/policy_reports/forest_conflict_2002.pdf Accessed on July 16, 2013.

Hazra, Sugata, "Climate Change Adaptation in Coastal Region of West Bengal", Climate Change Policy Paper II. Available at http://awsassets.wwfindia.org/downloads/climatechangeadaptation_in_coastal_regionof_west_bengal.pdf Accessed on July 26, 2013.

Hegde, N.G., "Challenges of Community Forestry in India", Asia Pacific Forestry Research – Vision 2010, Proceedings of the Regional Seminar, Kuala Lumpur, Malaysia, 2000. Available at www.fao.org/3/a-w7732e.pdf Accessed on December 21, 2014.

Hoffman, Justin, "The Maldives and Rising Sea Levels", ICE Case Studies, No. 206, May 2007. Available at http://www1.american.edu/ted/ice/maldives.htm Accessed on December 21, 2014.

ICLEI South Asia, *Renewable Energy and Energy Efficiency Status in India*, May 2007. Available at http://localrenewables.iclei.org/fileadmin/template/projects/localrenewables/files/Local_Renewables/Publications/REEE_report_India_final_sm.pdf Accessed on September 8, 2013.

IISD Reporting Services, "Summary of the First International Environment Forum for Basin Organizations", *Basin Organizations Forum Bulletin*, Vol. 227, No. 1, 2014, pp.1–13. Available at http://www.iisd.ca/download/pdf/sd/crsvol227num1e.pdf Accessed on March 21, 2015.

Impact on Vietnam Agriculture. Available at www.tiempocyberclimate.org/portal/archieve/vietnam/impact5.html Accessed on September 5, 2012.

In Farmers' Protests, a Climate Change Connection Lurks, December 10, 2020. Available at https://science.thewire.in/economy/agriculture/farmers-protest-climate-change-agriculture/ Accessed on January 23, 2022.

Ionesco, Dina, "Climate Migration: From the Paris Agreement to the Global Compact for Migration", *IOM, UN Migration Blog*, November 30, 2017. Available at https://weblog.iom.int/climate-migration-paris-agreement-global-compact-migration Accessed on January 20, 2022.

"Is JFM Relevant?", *Down to Earth*, September 15, 2011. Available at http://www.downtoearth.org.in/content/jfm-relevant Accessed on July 20, 2014.

Jacobs, Michael, "Lima Deal Represents a Fundamental Change in Global Climate Regime", *The Guardian*, December 15, 2014. Available at http://www.theguardian.com/environment/2014/dec/15/lima-deal-represents-a-fundamental-change-in-global-climate-regime Accessed on December, 17, 2014.

Jain, Ayesha, "Displacement Explained: How Many Climate Refugees Does India Have?", *The Quint,* August 11, 2021. Available at https://www.thequint.com/climate-change/explainer-who-are-climate-refugees-and-why-migration-is-rampant#read-more Accessed on January 22, 2021.

Jayaraman, T., "Climate Change and Agriculture: A Review Article with Special Reference to India", *Review of Agrarian Studies*, pp.16–78. Available at www.ras.org.in/climatechange_and_agriculture Accessed on July 10, 2014.

Johnson, Laura Story, "Environment, Security and Environmental Refugees", *Journal of Animal and Environmental Law*, Vol. 1, 2009–2010, pp.222–248. Available at https://drive.google.com/file/d/0B0gcImiUSq5Ebm9TZ1VwZlZzd2s/view?pli=1 Accessed on November 16, 2012.

Khor, Martin, "Rich Nations Plot Kyoto's End", *The Star*, October 5, 2009. Available at http://www.twnside.org.sg/title2/climate/info.service/2009/climate. Change.20091001.htm Accessed on April 4, 2013.

Kolmannskog, Vikram, "Climate Change-Related Displacement and the European Response", Paper presented at SID Vijverberg Session on Climate Change and Migration, The Hague, January 20, 2009. Available at sideurope.files.wordpress.com/2009/02/presentation-kolmannskog.doc Accessed on June 17, 2014.

Krishnan, N. R., "Doha Climate Talks Not a Failure", *The Hindu*, January 2, 2013. Available at http://www.thehindubusinessline.com/opinion/doha-clima... Accessed on April 13, 2014.

Krishnan, N. R., "The Climate Turned Against India at Durban", *The Hindu*, December 13, 2012. Available at http://www.thehindubusinessline.com/opinion/the-climate-turned-against-india-at-durban/article2709519.ece Accessed on January 13, 2013.

Kumar Sambhav S., "How Effective Is India's Disaster Management Authority?", *Down to Earth*, June 20, 2013. Available at http://www.downtoearth.org.in/content/how-effective-indias-disaster-management-authority Accessed on August 1, 2014.

Kumaraperumal, R., S. Natarajan, S. Chellamuthu, S. S. Ganesh and G. Anandakumar, "Impact of Tsunami 2004 in Coastal Villages of Nagapattinam District, India", *Science of Tsunami Hazards*, Vol. 26, No. 2, 2007, pp.93–114. Available at http://tsunamisociety.org/262Kumarap.pdf Accessed on January 3, 2015.

Lama, Mahendra P., "Internal Displacement in India: Causes, Protection and Dilemmas", *Forced Migration Review*, Vol. 8, pp.24–26. Available at http://www.fmreview.org/FMRpdfs/FMR08/fmr8.9.pdf Accessed on August 28, 2013.

Macfarquhar, Neil, "U.N. Deadlock on Addressing Climate Shift", *New York Times*, July 20, 2011. Available at http://www.nytimes.com/2011/07/21/world/21natio ns.html?_r=0 Accessed on July 8, 2014.

McAdam, Jane, "Climate Change Displacement and International Law: Complementary Protection Standards", Legal and Protection Policy Series, *UNHCR, Division of International Protection*, May 2011. Available at http://www.unhcr.org/4dff16e99.pdf Accessed on June 18, 2014.

Michel, David and Amit Pandya (eds.), *Troubled Waters: Climate Change, Hydropolitics, and Transboundary Resources*, The Henry L. Stimson Center: Washington, DC, 2009. Available at www.stimson.org/images/uploads/.../Troubled_Waters-Complete.pdf Accessed on June 13, 2013.

Mohanty, Avinash, "Preparing India for Exteme Climate Events", *Hindustan Times*, July 1, 2021. Available at https://www.hindustantimes.com/ht-insight/climate-change/preparing-india-for-extreme-climate-events-101625127345593.html Accessed on January 19, 2021.

Mora, Gabriella M. Cebada, *Reduced Emission from Deforestation and Forest Degradation*, Carbon Count Series, Antioch University: New England. Available at www.antiochne.edu/wp-content/uploads/2012/.../REDDpresentation.pdf Accessed on November 11, 2013.

Natarajan, Jayanti, *Suo Moto Statement in Lok Sabha by Minister of State for Environment and Forests on Durban Agreements*, December 16, 2011. Available at pib.nic.in/newsite/erelease.aspx?relid=78811 Accessed on January 27, 2014.

"National Mission on Clean Coal Technologies on Cards", *The Hindu*, February 27, 2012. Available at http://www.thehindubusinessline.com/economy/

national-mission-on-clean-coal-technologies-on-cards/article2938979.ece Accessed on March 21, 2015.

Nayak, Avaya K., "Post Super Cyclone Orissa: An Overview", *Orissa Review*, October 2009. Available at http://orissa.gov.in/e-magazine/Orissareview/2009/October/engpdf/Pages98–104.pdf Accessed on July 28, 2013.

O'Brien Karen L. and Robin M. Leichenko, *Climate Change, Globalization and Water Scarcity*, 2008. Available at http://www.google.co.in/url?sa=t&rct=j&q=&esrc=s&source=web&cd=1&ved=0CCEQFjAA&url=http%3A%2F%2Fwww.zaragoza.es%2Fcontenidos%2Fmedioambiente%2FcajaAzul%2F17S6-P2-ObrienACC.pdf&ei=ZwwCVJffItWVuATi2YLIAw&usg=AFQj CNEbO6WuzpsbMp22SgxgWBK gFY9mmg&bvm=bv.74115972,d.c2E Accessed on August 29, 2014.

Parry, Martin, Alex Evans, Mark W. Rosegrant and Tim Wheeler, "Climate Change and Hunger: Responding to the Challenge", *World Food Programme*, May 2009. Available at http://www.wfp.org/content/climate-change-and-hunger-responding-challenge Accessed on March 19, 2015.

Pearce, Fred, "Climate Change Is Not an Excuse for Genocide", *The Telegraph*, January 8, 2008. Available at http://www.telegraph.co.uk/news/earth/earthcomment/3320845/Climate-change-is-not-an-excuse-for-genocide.html Accessed on November 13, 2013.

Piontek, Franziska, P. Michael Link and Jurgen Scheffran, *Impacts of Climate Change on the Nile River Conflict: The Case of Egypt*. Available at http://clisec.zmaw.de/fileadmin/user.../piontek-et-al-2010_amman.pdfa Accessed on June 25, 2012.

"Protect Papua New Guinea's Rainforest: COP26 REDD+Carbon Credits Auction", *Environmental Finance*, October 18, 2021. Available at environmental-finance.com Accessed on January 20, 2022.

PTI, "14,000 sq.km. Land at Risk with Rising Sea Level", *The Hindu*, June 18, 2013. Available at http://www.thehindubusinessline.com/news/14000-sqkm-land-at-risk-with-rising-sea-level-report/article4826559.ece Accessed on July 24, 2013.

Rai, Dipu and Mayank Mishra, "Rising Frequency and Intensity of Flood, Drought and Cyclone", *India Today*, June 12, 2021. Available at indiatoday.in Accessed on December 19, 2021.

"Rains in Uttarakhand Extreme Weather Event: Scientists", *Down to Earth*, June 18, 2013. Available at http://www.do wntoearth.org.in/content/rains-over-past-three-days-extreme-weather-event-scientists Accessed on March 2, 2015.

Rajamani, Lavanya, "The Can-Can't at Cancun", *Indian Express*, December 14, 2010, http://www.indianexpress.com/story-print/724374/ Accessed on January 14, 2014.

"Recommendations for Short-Term Actions", Policy Brief, International Institute for Sustainable Development, August 2013. Available at http://www.iisd.org/gsi/sites/default/files/pb16_prioritizing.pdf Accessed on June 26, 2014.

"Report of the Fourth Conference of the Parties to the United nations Framework Convention On Climate Change", *Earth Negotiations Bulletin, International Institute for Sustainable Development*, Vol. 12, No. 97, November 16, 1998, pp.3–13.

Revi, Aromar, "Climate Change Risk: An Adaptation and Mitigation Agenda for Indian Cities", *Environment and Urbanization*, Vol. 20, No. 1, April 2008, pp.207–229. Available at http://pubs.iied.org/pdfs/G02275.pdf Accessed on January 3, 2015.

"Rise in Sea Level Threatening Sundarbans", *The Hindu*, January 6, 2013. Available at http://www.thehindubusinessline.com/news/science/rise-in-sea-level-threatening-sundarbans-sayspachauri/article4279603.ece?ref=relatedNews Accessed on July 26, 2013.

Roman, Yavich, "Waxman-Markey and Failed Senate Legislation: Climate Change Policy Case Study", *Environmental and Natural Resource Policy*, December 2010, pp.1–17. Available at cepa.maxwell.syr.edu/pages/206/WMClimateChange.pdf Accessed on April 29, 2013.

Sadoff, Claudia and Mike Muller, "Water Management, Water Security and Climate Change Adaptation: Early Impacts and Essential Responses", Global Water Partnership Technical Committee Background Paper, No. 14, 2009. Available at www.gwp.org/Global/GWP...Files/.../tec14.pd Accessed on May 27, 2014.

Sarabhai, Kartikeya, *What Are the Threats to Biodiversity*. Available at http://www.Vigyanprasar.gov.in/Radioserials/Threat_to_biodiversity_-_draft_paper.pdf Accessed on November 15, 2014.

Saran, Shyam Sara, *India's Climate Change Policy: Towards a Better Future*, November 8, 2019. Available at https://www.mea.gov.in/articles-in-indian-media.htm?dtl/32018/Indias_Climate_Change_Policy_Towards_a_Better_Future Accessed on January 21, 2022.

Saran, Shyam, India's *Climate Change Strategy after Durban, Occasional Paper, Perspectives in Indian Development*, New Series 1, Nehru Memorial Museum and Library: New Delhi, 2012.

Second Largest Rate of Amazon Deforestation in Brazilian History, Greenpeace Press Release, May 19, 2005. Available at http://www.greenpeace.org/international/en/press/releases/second-largest-rate-of-amazon/ Accessed on December 21, 2014.

Seth, Bharat Lal, "National Water Policy, 2012 Silent on Priorities", *Down to Earth*, February 10, 2012. Available at http://www.downtoearth.org.in/content/national-water-policy-2012-silent-priorities Accessed on July 19, 2014.

Sharma, J. V. and Priyanka Kohli, "Forest Governance and Implementation of REDD+ in India", Policy Brief, Ministry of Environment and Forest, Government of India and Tata Energy and Resources Institute. Available at indiaenvironmentportal.org.in/.../forest-governance-and-implementation... Accessed on July 20, 2014.

Shrivastava, Kumar Sambhav, "Green Tribunal Spells Its Mandate", *Down to Earth*, January 1–15, 2013, p.16. Available at http://www.downtoearth.org.in/content/green-tribunal-spells-its-mandate Accessed on March 21, 2015.

Somanathan, E., *Biodiversity in India, Prepared for the Oxford Companion to Economics in India*. Available at www.isid.ac.in/~som/papers/BiodiversityinIndia_rev.pdf Accessed on November 15, 2014.

Sud, Ridhima, Jitendra Vir Sharma, Arun Kumar Bansal and Subhash Chandra, "Institutional Framework for Implementing REDD+ in India", Ministry of Environment and Forest, Government of India, The Energy and Resources Institute, Norwegian Embassy, 2012, Available at www.indiaenvironmentportal.org.in/.../institutional-framework-for-imple... Accessed on July 20, 2014.

Sunita, Narain, "India's New Climate Targets: Bold, Ambitious and a Challenge for the World", *Down to Earth*, November 2, 2021. Available at downtoearth.org.in Accessed on January 12, 2022.

Sustainable Agriculture: No to GMOs. Available at http://www.greenpeace.org/usa/en/campaigns/genetic-engineering/ Accessed on July 14, 2014.

Taenzler, Dennis, Lukas Ruettinger, Katherina Ziegenhagen (adelphi) and Gopalakrishna Murthy, Water, *Crisis and Climate Change in India: A Policy Brief, The Initiative for Peacebuilding – Early Warning Analysis to Action (IfP-EW)*, October, 2011. Available at www.ifp-ew.eu/pdf/201110IfPEWWater CrisisIndiaPolBrf.pdf Accessed on 18 July, 2014.

Thakkar, Himanshu, *Water Sector Options for India in a Changing Climate*, South Asia Network on Dams, Rivers & People: Delhi, March, 2012. Available at sandpr.in/wtrsect/Water_Sector_options_In Accessed on March 19, 2015.

Timperley, Jocelyn, "The Broken $100-Billion Promise of Climate Finance – and How to Fix It", *Nature*, October 20, 2021. Available at https://www.nature.com/articles/d41586-021-02846-3 Accessed on January 23, 2022.

Tisdell, Clem, "Global Warming and the Future of Pacific Island Countries", Working Paper No. 147, The University of Queens Land, 2007. Available at http://www. uq.edu.au/rsmg/docs/ClemWPapers/EEE/WP147.pdf Accessed on December 20, 2014.

Tiseo, Ian, "Carbon-dioxide Emissions from Fossil Fuel Combustion in India from 1960 to 2020, by Fuel Type", *Statista*, January 18, 2022. Available at Accessed on January 16, 2021.

Two Very Different Views on the Warsaw REDD Deal from Indigenous Peoples Organisations. Available at http://www.redd-monitor.org/2013/12/07/two-very-different-views-on-the-war saw-red-deal-from-indigenous-peoples-organisations/ Accessed on June 23, 2014.

UN News Service, *The Future Starts Now,' Ban Says at Launch of UN Decade of Sustainable Energy for All*, June 5, 2014. Available at http://www.un.org/apps/news/story.asp?NewsID= 47969 #.U6 z7ytdDu8Y Accessed on June 27, 2014.

Uthra, Radhakrishnan, "Power Games at UN Climate Talks", *Down to Earth*, December 15, 2013. Available at http://www.downtoearth.org.in/content/power-games-un-climate-talks Accessed on March 21, 2015.

Wamukoya, George, *COP 15 Outcomes for Reducing Emissions from Agriculture and Other Land Uses*, 2010. Available at http://www.iisd.org/pdf/2010/03REDD_II_Hue_Agriculture.pdf Accessed on May 19, 2014.

Werrell, Caitlin E. and Francesco Femia, *The Arab Spring and Climate Change: A Climate and Security Correlations Series,* Center for American Progress, STIMSON, The Center for Climate and Security, February 2013. Available at https://climateandsecurity.files.wordpress.com/2012/04/climatechangearabspring-ccs-cap-stimson.pdf Accessed on December 14, 2014.

Whitley, Shelagh, *Time to Change the Game: Fossil Fuel Subsidies and Climate*, Overseas Development Institute, November 2013. Available at www.odi.org.uk/sites/odi.../8669.pdf Accessed on June 25, 2014; www.wfp.org/content/climate-change-and-hunger-responding-challenge Accessed on July 10, 2010.

Zafar-ul Islam, M., Shaily Menon, Xingong Li and A. Townsend Peterson, "Forecasting Ecological Impacts of Sea-Level Rise on Coastal Conservation Areas in India", *Journal of Threatened Taxa*, Vol. 5, No. 9, May 26, 2013, pp.4349–4358. Available at http://dx.doi.org/10.11609/JoTT.o3163.4349–58 Accessed on July 24, 2013.

Index